MATH MASTER 1

MATH MASTER 1

Strategies for Computation and Problem Solving

Jerry Howett

CAMBRIDGE Adult Education
Prentice Hall Regents, Englewood Cliffs, NJ 07632

Executive Editor: JAMES W. BROWN
Editorial supervision: TIM FOOTE
Production supervision and interior design: TUNDE A. DEWEY
Manufacturing buyer: MIKE WOERNER
Photo Research: PAGE POORE
Cover Photo: Image Bank/GARRY GAY

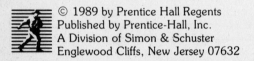
© 1989 by Prentice Hall Regents
Published by Prentice-Hall, Inc.
A Division of Simon & Schuster
Englewood Cliffs, New Jersey 07632

All rights reserved. No part of this book may be
reproduced, in any form or by any means,
without permission in writing from the publisher.

Printed in the United States of America
10 9 8 7 6 5 4 3 2

ISBN 0-13-943960-9

Prentice-Hall International (UK) Limited, *London*
Prentice-Hall of Australia Pty. Limited, *Sydney*
Prentice-Hall Canada Inc., *Toronto*
Prentice-Hall Hispanoamericana, S.A., *Mexico*
Prentice-Hall of India Private Limited, *New Delhi*
Prentice-Hall of Japan, Inc., *Tokyo*
Simon & Schuster Asia Pte. Ltd., *Singapore*
Editora Prentice-Hall do Brasil, Ltda., *Rio de Janeiro*

Acknowledgement

The author wishes to thank the following for their contributions during the preparation of this book.

Sharon Conrick

Washington-Warren-Hamilton-Essex B.O.C.E.S.
Hudson Falls, New York

Ellen T. DeSantis

Maryland Correctional Institution—Jessup
Jessup, Maryland

Lynne Ornstein

The Fortune Society
New York, New York

CONTENTS

To the Teacher   xi

Part A   WHOLE NUMBERS   1

Whole Numbers Pretest   2

Whole Numbers Pretest Record   3
Whole Numbers Pretest Lesson Guide   4

Chapter 1   Addition   5

Lesson 1   Introducing Whole Numbers   5
Lesson 2   Addition Facts   7
Lesson 3   Adding with No Carrying   10
Lesson 4   Carrying   14
Lesson 5   Using a Calculator   19
Lesson 6   Long Column Addition   21
Lesson 7   Reading and Writing Whole Numbers   22
Lesson 8   Rounding Off   25
Lesson 9   Addition Word Problems   27
Lesson 10   Bank Charges   30
Lesson 11   Using a Menu   32
Lesson 12   Bar Graphs   34

Chapter 2   Subtraction   37

Lesson 13   Subtraction Facts   37
Lesson 14   Subtracting with No Borrowing   39
Lesson 15   Borrowing   41
Lesson 16   Borrowing across Zeros   44
Lesson 17   Mixed Subtraction Problems   47

Lesson 18 Subtraction Word Problems 48
Lesson 19 Mixed Addition
          and Subtraction Word Problems 51
Lesson 20 Balancing a Checkbook 54
Lesson 21 Tables 55
Lesson 22 Bar Graphs 56

Chapter 3   Multiplication   58

Lesson 23 Multiplication Facts 58
Lesson 24 Multiplying by One Digit with No
          Carrying 61
Lesson 25 Multiplying by One Digit with
          Carrying 63
Lesson 26 Multiplying by Two Digits 65
Lesson 27 Multiplying by Three Digits 68
Lesson 28 Multiplying by 10, 100, and 1000 70
Lesson 29 Multiplication Word Problems 71
Lesson 30 Mixed Operation Word Problems 74
Lesson 31 The Distance Formula 76
Lesson 32 Perimeter of a Rectangle 78
Lesson 33 Area of a Rectangle 80
Lesson 34 Volume of a Rectangular Solid 82
Lesson 35 Tables 84
Lesson 36 Bar Graphs 86

Chapter 4   Division   88

Lesson 37 Division Facts 88
Lesson 38 Dividing by One Digit 90
Lesson 39 Remainders 93
Lesson 40 Zeros as Place Holders 94
Lesson 41 Dividing by Two Digits 95
Lesson 42 Dividing by Three Digits 97
Lesson 43 Dividing by 10, 100, and 1000 99
Lesson 44 Division Word Problems 101
Lesson 45 Mixed Multiplication
          and Division Word Problems 104
Lesson 46 Average 106
Lesson 47 Unit Prices 108
Lesson 48 Converting Temperatures 109

Whole Numbers Review   111

Whole Numbers Review Record   115
Whole Numbers Review Lesson Guide   115

Part B   DECIMALS   117

Decimals Pretest   118

Decimals Pretest Record   119
Decimals Pretest Lesson Guide   119

Chapter 1   Addition and Subtraction   120

Lesson 1   Introducing Decimals   120
Lesson 2   Adding Decimals   123
Lesson 3   Reading and Writing Decimals   126
Lesson 4   Comparing Decimals   129
Lesson 5   Subtracting Decimals   130
Lesson 6   Mixed Addition and Subtraction   132
Lesson 7   Mixed Addition and Subtraction Word Problems   134
Lesson 8   Multistep Word Problems   136
Lesson 9   Tables   138
Lesson 10   Bar Graphs   139
Lesson 11   Reading a Metric Ruler   141

Chapter 2   Multiplication   143

Lesson 12   Multiplying Decimals   143
Lesson 13   Multiplying by 10, 100, and 1000   147
Lesson 14   Rounding Off Decimals   148
Lesson 15   Multiplication Word Problems   150
Lesson 16   Mixed Operation Word Problems   152
Lesson 17   Perimeter and Area of Rectangles   154
Lesson 18   Circles   155
Lesson 19   Circumference of a Circle   157
Lesson 20   Area of a Circle   158

Chapter 3 Division 160

Lesson 21 Dividing Decimals by Whole Numbers 160
Lesson 22 Dividing Decimals by Decimals 162
Lesson 23 Dividing Whole Numbers by Decimals 165
Lesson 24 Mixed Division Problems 167
Lesson 25 Uneven Decimal Division 168
Lesson 26 Dividing by 10, 100, and 1000 170
Lesson 27 Division Word Problems 171
Lesson 28 Mixed Operation Word Problems 173
Lesson 29 Average 175
Lesson 30 Gas Mileage 176
Lesson 31 Batting Averages 178

Decimals Review 180

Decimals Review Record 184
Decimals Review Lesson Guide 184

ANSWERS 185

TO THE TEACHER

Math Master 1 and *Math Master 2* are text/workbooks for adults who want to refresh or develop arithmetic skills. The two books present a course in four parts. *Math Master 1* covers whole numbers and decimals. Fractions and percents, the other two parts, are covered in *Math Master 2*.

Throughout the course, adults are shown common-sense approaches to solving problems. When there is a relatively easy way to solve a problem, that way is taught rather than a more complicated one. The course teaches adults how to compute both by hand and on a calculator. The word problems in each part of the course are accompanied by questions that encourage thinking each problem through logically to its solution. Each part of the course contains lessons that show an adult how to apply the skill taught to the kinds of problems he or she encounters in daily life. The following paragraphs tell, in more detail, how the material in each of the four parts of the course is organized.

A part begins with a short diagnostic **Pretest** that identifies an adult's strengths and weaknesses at pencil-and-paper calculations. A **Pretest Record** tells him or her which lessons to study in the part. In addition, a **Pretest Lesson Guide** tells which lesson addresses each problem type in the Pretest.

There are three to five chapters in each part of the course. Each chapter opens with lessons that provide step-by-step instruction in the operation at hand—for example, the addition of whole numbers. Instruction includes explanations, examples, hints, and illustrations. It also shows how to perform operations on a calculator and encourages the use of calculators as one way to check answers. Extensive exercise in each lesson helps students learn skills solidly.

Chapters continue with lessons on word problems set in adult contexts. To help students employ sound problem-solving techniques, key operations words are highlighted. In addition, problems are followed by questions that help students realize how to reach solutions. Mixed-operations word problems are a regular feature in each part of the course.

Several lessons follow that concentrate on application of the chapter's skills in situations that are typical in adults' work and personal lives. For example, lessons on the formulas for distance, perimeter, area, and volume are among those that end the chapter on whole number multiplication.

At the end of each part of the course there is a comprehensive **Review**. Adults can record their scores in the **Review Record** and use the **Review Lesson Guide** to find which lessons to go over, if any.

All in all, the course is designed to make math masters of the adults who take it and apply what they learn in their daily lives.

MATH MASTER 1

**Strategies for Computation
and Problem Solving**

PART A

WHOLE NUMBERS

Whole Numbers Pretest

Chapter 1 Addition
Chapter 2 Subtraction
Chapter 3 Multiplication
Chapter 4 Division

Whole Numbers Review

Whole Numbers Pretest

These problems will help you decide what you need to study most about whole numbers. Solve every problem that you can. Then check your answers. Finally, fill in the Whole Numbers Pretest Record that follows the Pretest. It will show you which pages you should study.

Addition

1. 426 + 532 =

2. 2,407 + 1,392 =

3. 89 + 76 =

4. 473 + 986 =

5. 12,942 + 8,073 =

6. 186,248 + 39,732 =

7. 32
    47
    93
    78
   +56

8.   58
    126
     34
    236
   + 91

9. 2,056 + 13,983 + 218 =

10. 4,926 + 377 + 10,284 =

Subtraction

11. 423 − 212 =

12. 1690 − 1240 =

13. 93 − 27 =

14. 156 − 88 =

15. 6,248 − 2,919 =

16. 12,036 − 9,448 =

17. 20,000 − 16,273 =

18. 146,203 − 95,862 =

19. 104,260 − 92,055 =

20. 760,044 − 29,756 =

Multiplication

21. 43 × 2 =

22. 285 × 6 =

23. 3,609 × 4 =

24. 7,850 × 8 =

25. 40,206 × 7 =

26. 675 × 93 =

27. 32,298 × 12 =

28. 685 × 430 =

29. 493 × 126 =

30. 706 × 1000 =

Division

31. 140 ÷ 5 =

32. 1296 ÷ 6 =

33. 4832 ÷ 8 =

34. 6440 ÷ 7 =

35. 252 ÷ 14 =

36. 828 ÷ 36 =

37. 3312 ÷ 92 =

38. 2295 ÷ 27 =

39. 1944 ÷ 216 =

40. 1300 ÷ 100 =

Check your answers on page 185.

WHOLE NUMBERS PRETEST RECORD

Section	Circle the numbers of the problems you got right	If your score is	Study pages
Addition	1 2 3 4 5 6 7 8 9 10	0 - 7 8 - 10	5 - 36 27 - 36
Subtraction	11 12 13 14 15 16 17 18 19 20	0 - 7 8 - 10	37 - 57 48 - 57
Multiplication	21 22 23 24 25 26 27 28 29 30	0 - 7 8 - 10	58 - 87 70 - 87
Division	31 32 33 34 35 36 37 38 39 40	0 - 7 8 - 10	88 - 110 101 - 110

For each problem in the Whole Numbers Pretest, the lesson that teaches about that kind of problem is listed in this guide.

WHOLE NUMBERS PRETEST LESSON GUIDE

Problem Number	1	2	3	4	5	6	7	8	9	10	11	12
Lesson Number	3	3	4	4	4	4	6	6	4	4	14	14
Problem Number	13	14	15	16	17	18	19	20	21	22	23	24
Lesson Number	15	15	15	16	16	16	16	16	24	25	25	25
Problem Number	25	26	27	28	29	30	31	32	33	34	35	36
Lesson Number	25	26	26	27	27	28	38	38	40	40	41	41
Problem Number	37	38	39	40								
Lesson Number	41	41	42	43								

Whole Numbers

CHAPTER 1: Addition

Lesson 1

INTRODUCING WHOLE NUMBERS

DIGITS

Whole numbers are written with the **digits** 0, 1, 2, 3, 4, 5, 6, 7, 8, and 9.
  The number 144 has three digits.
  The number 2500 has four digits.

PLACE VALUE

Each position in a whole number has a **place value**. The name of each place is a clue to the value of each digit. Below is a diagram of the first four whole-number places and their values.

As you move left in the number system, each place gets ten times bigger. The units place has the value 1. The tens have a value of $10 \times 1 = 10$. (Read $10 \times 1 = 10$ as, "Ten times one equals ten.") The hundreds have a value of $10 \times 10 = 100$. The thousands have a value of $10 \times 100 = 1000$.

Place Names	thousands	hundreds	tens	units or ones
Places	___	___	___	___
Place Values	1000	100	10	1

Think about the number **4978**. Its place values are shown in the drawing.

- 8 is in the units, or ones, place. 8 has a value of $8 \times 1 = 8$.
- 7 is in the tens place. 7 has a value of $7 \times 10 = 70$.
- 9 is in the hundreds place. 9 has a value of $9 \times 100 = 900$.
- 4 is in the thousands place. 4 has a value of $4 \times 1000 = 4000$.

Place Names	thousands	hundreds	tens	units or ones
Places	4	9	7	8
Place Values	1000	100	10	1

ZEROS

Watch for zeros. The digit 0 has no **value**, but it has an important **use** as a placeholder. In the number 605, the digit 5 is in the ones place. 5 has a value of $5 \times 1 = 5$.

The zero in the tens place has a value of 0 × 10 = 0. The zero has no value, but it **holds** the tens place. And it puts 6 in the hundreds place. 6 has a value of 6 × 100 = 600.

EXERCISE

1. Circle the two-digit numbers.
   14   327   60   1580   45   700

2. Circle the three-digit numbers.
   2486   27   685   93,000   450   700

3. Circle the four-digit numbers.
   3000   12,567   1630   1988   456   150,000

4. Circle the digit in the ones place in each number.
   287   16   1340   77   489   61

5. Circle the digit in the tens place in each number.
   345   18   2580   418   26   1492

6. Circle the digit in the hundreds place in each number.
   1250   923   7789   2344   800   1620

7. Circle the digit in the thousands place in each number.
   4500   1878   1925   6650   2399   3776

Use the number **3812** to answer questions 8 to 11. Fill in the blanks.

8. 2 is in the _____ place. 2 has a value of _____
9. 1 is in the _____ place. 1 has a value of _____
10. 8 is in the _____ place. 8 has a value of _____
11. 3 is in the _____ place. 3 has a value of _____

Use the number **6050** to answer questions 12 and 13. Fill in the blanks.

12. 5 is in the _____ place. 5 has a value of _____
13. 6 is in the _____ place. 6 has a value of _____

Answers are on page 185.

Lesson 2

ADDITION FACTS

EXERCISE

This exercise gives you practice with the addition facts. Write every answer that you know. Use the guide that follows this exercise to learn any facts you do not already know.

1. 2   4   1   7   9   6   1   9   4   8
   +6   +8   +7   +3   +8   +6   +5   +0   +7   +3

2. 3   0   9   6   8   7   4   6   3   2
   +5   +8   +4   +5   +1   +9   +2   +9   +3   +7

3. 6   2   6   3   7   9   1   2   8   6
   +7   +5   +3   +1   +4   +7   +9   +1   +7   +1

4. 3   5   4   7   8   0   2   5   8   3
   +9   +7   +6   +1   +6   +7   +9   +0   +4   +2

5. 8   3   9   0   6   5   9   1   2   5
   +8   +4   +6   +3   +2   +9   +3   +1   +8   +4

6. 0   7   1   5   7   0   1   5   4   2
   +9   +2   +6   +5   +8   +1   +2   +6   +9   +4

7. 2   4   8   2   9   8   1   5   7   9
   +3   +0   +9   +2   +1   +5   +3   +3   +6   +5

8. 3   7   9   4   3   4   9   6   1   7
   +7   +8   +2   +5   +6   +1   +9   +4   +8   +7

9. 5   3   4   7   5   8   0   7   6   4
   +1   +8   +4   +0   +2   +2   +6   +5   +8   +3

Answers are on page 185.

Chapter 1: Addition 7

ADDITION FACTS GUIDE

Try learning a few addition facts at a time.

Facts That Add to 10

The facts in the next row all add up to 10.

1	2	3	4	5	6	7	8	9
+9	+8	+7	+6	+5	+4	+3	+2	+1
10	10	10	10	10	10	10	10	10

Facts with 9

When you add 9 to a number, the sum is one less than that number plus 10. For example, when 9 is added to 5, the sum is 14.

one less than 5	4
plus 10	+10
sum	14

The following are the addition facts with 9.

1	2	3	4	5	6	7	8	9
+9	+9	+9	+9	+9	+9	+9	+9	+9
10	11	12	13	14	15	16	17	18

Facts That Add to More Than 10

Below are the addition facts that add up to more than 10.

2
+9
11

3	3
+8	+9
11	12

4	4	4
+7	+8	+9
11	12	13

5	5	5	5
+6	+7	+8	+9
11	12	13	14

8　Whole Numbers

6	6	6	6	6			
+5	+6	+7	+8	+9			
11	12	13	14	15			
7	7	7	7	7	7		
+4	+5	+6	+7	+8	+9		
11	12	13	14	15	16		
8	8	8	8	8	8	8	
+3	+4	+5	+6	+7	+8	+9	
11	12	13	14	15	16	17	
9	9	9	9	9	9	9	9
+2	+3	+4	+5	+6	+7	+8	+9
11	12	13	14	15	16	17	18

As you work through this book, look back at this guide until you have memorized every addition fact here.

Lesson 3

ADDING WITH NO CARRYING

ADDING

The answer to an addition problem is the **sum** or **total**. To add numbers, line up the numbers with units under units, tens under tens, and so on. Begin adding with the units.

EXAMPLE: 23
          +56

```
   1              2
 tens units    tens units
   23             23
  +56            +56
    9             79
```

Step 1. Add the digits in the units column. 3 + 6 = 9. Write 9 in the units place in the sum.

Step 2. Add the digits in the tens column. 2 + 5 = 7. Write 7 in the tens place in the sum. The sum is 79.

CHECKING

To check a sum, add the numbers from bottom to top. Again, start with the units. The new sum should be the same as the first sum.

```
   Adding        Checking
                    79
    23              23
   +56             +56
    79              79
```

10   *Whole Numbers*

EXERCISE A

Add and check each problem. When you add, write the first sum below the problem. When you check, write the new sum in the box above the problem.

1.
□	□	□	□	□	□
25	53	67	54	83	29
+43	+26	+30	+25	+12	+50

2.
□	□	□	□	□	□
77	18	24	66	25	40
+21	+31	+73	+23	+61	+58

3.
□	□	□	□	□
805	631	203	430	524
+183	+246	+530	+264	+404

4.
□	□	□	□	□
437	126	388	559	276
+241	+352	+311	+240	+623

5.
□	□	□	□
1,270	4,563	2,781	7,063
+5,129	+2,414	+4,115	+2,824

6.
□	□	□	□
23,045	32,914	73,502	40,851
+61,522	+41,055	+21,493	+36,142

Answers are on page 185.

REWRITING HORIZONTAL PROBLEMS

When an addition problem is written horizontally (with the numbers side by side), line up the numbers with units under units, tens under tens, and so on.

EXAMPLE: 325 + 71 =

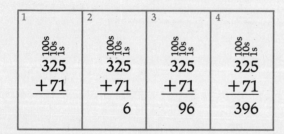

Step 1. Line up the numbers with units under units, and tens under tens.
Step 2. Add the digits in the units column. 5 + 1 = 6.
Step 3. Add the digits in the tens column. 2 + 7 = 9.
Step 4. Add the digits in the hundreds column. 3 is the only digit in the hundreds column. 3 + nothing = 3. The sum is 396.

EXERCISE B

Add and check each problem.

1. 42 + 53 =   54 + 22 =   61 + 36 =   82 + 16 =

2. 13 + 70 =   28 + 41 =   50 + 38 =   83 + 14 =

3. 412 + 53 =   35 + 41 =   878 + 121 =   69 + 120 =

4. 128 + 30 =   392 + 206 =   816 + 52 =   225 + 34 =

5. 4,332 + 2,567 =   8,015 + 1,940 =   4,660 + 3,015 =

12 Whole Numbers

6. 4,126 + 421 =   3,802 + 1,056 =   46,237 + 2,502 =

7. 10,344 + 5,325 =   25,267 + 142,322 =   30,524 + 9,143 =

Answers are on page 185.

Lesson 4

CARRYING

TWO-DIGIT SUMS

The sum of the digits in one column is often a two-digit number. Look at the problem below. The sum of 8 and 5 is a two-digit number, 13. Write the 3 in the units column, and the 1 in the tens column.

$$\begin{array}{r} 8 \\ +5 \\ \hline 13 \end{array}$$

ADDING TWO COLUMNS OR MORE

When you are adding two or more columns, you may have to *carry* a digit from one column to the next.

EXAMPLE: 78
         +54

1	2
tens units	tens units
1	1
78	78
+54	+54
2	132

Step 1. Add the digits in the units column. 8 + 4 = 12. Write the 2 in the units place, and **carry** the 1 to the tens column. The digit 1 is 1 ten.

Step 2. Add the digits in the tens column, beginning with the digit you carried. 1 + 7 = 8 and 8 + 5 = 13. Write the 3 in the tens column, and the 1 in the hundreds column. The sum is 132.

14   Whole Numbers

Exercise A

Add and check each problem.

1.
47	56	83	27	59	68
+64	+99	+28	+74	+82	+35

2.
81	29	55	48	66	91
+39	+27	+88	+92	+75	+59

3.
268	473	534	297	356
+492	+577	+656	+378	+295

4.
912	36	788	92	647
+98	+477	+ 38	+218	+ 76

5.
1,756	4,983	2,656	4,353
+2,154	+4,109	+8,337	+5,569

6.
12,285	64,093	83,075	60,924
+25,946	+39,428	+14,716	+58,896

Answers are on page 185.

REWRITING HORIZONTAL PROBLEMS

Before you add a problem that is written horizontally (with the numbers side by side), rewrite the problem so that units are lined up under units, tens under tens, and so on.

EXAMPLE: 845 + 97 =

1	2	3	4
100s 10s 1s	100s 10s 1s 1	100s 10s 1s 1	100s 10s 1s 1
845 + 97	845 + 97 ――― 2	845 + 97 ――― 42	845 + 97 ――― 942

Step 1. Line up the numbers with units under units and tens under tens.

Step 2. Add the digits in the units column. 5 + 7 = 12. Write the 2 in the units place, and *carry* the 1 to the tens column. The digit 1 is 1 ten.

Step 3. Add the digits in the tens column, beginning with the digit you carried. 1 + 4 = 5 and 5 + 9 = 14. Write the 4 in the tens place, and *carry* the 1 to the hundreds column. The digit 1 is 1 hundred.

Step 4. Add the digits in the hundreds column, beginning with the digit you carried. 1 + 8 = 9. The sum is 942.

EXERCISE B

Add and check each problem.

1. 258 + 73 =   86 + 99 =   36 + 476 =

2. 66 + 1297 =   3055 + 938 =   528 + 4688 =

3. 3208 + 95 =   396 + 7804 =   2961 + 379 =

4. 427 + 36 =   1095 + 18 =   3446 + 278 =

5. 385 + 1882 =   216 + 7496 =   2937 + 84 =

Answers are on page 185.

ADDING MORE THAN TWO NUMBERS

To add more than two numbers, find the sum for each column. You can add the digits in a column in any order.

Example: 197 + 144 + 66 =

```
   1                  2                   3
  100s               100s                100s
   10s                10s                 10s
    1s                 1s                  1s
                      2 1                 2 1
   197                197                 197
   144                144                 144
 +  66              +  66               +  66
  ────               ────                ────
     7                 07                 407
```

Step 1. Set the problem up. Then add the digits in the units column. 7 + 4 = 11. Then 11 + 6 = 17. You could also start with 4 + 6 = 10. Then 10 + 7 = 17. Some people like to work with sums that add up to 10. Write the 7 in the units place, and carry the 1 to the tens column.

Step 2. Add the digits in the tens column. You can begin with the digit you carried. 1 + 9 = 10. Then 10 + 4 = 14. Then 14 + 6 = 20. Write the 0 in the tens column and carry the 2 to the hundreds column.

Step 3. Add the digits in the hundreds column. You can begin with the digit you carried. 2 + 1 = 3. Then 3 + 1 = 4. Write the four in the hundreds place. The sum is 407.

Exercise C

Add and check each problem.

1. 349 + 163 + 87 =   905 + 48 + 676 =

2. 88 + 1327 + 741 =   7233 + 980 + 204 =

3. 594 + 7 + 10,308 =   539 + 13,854 + 36 =

4. 2,035 + 817 + 33,256 =   29 + 718 + 4343 =

5. 115,206 + 1,412 + 73 =   1,203 + 109,562 + 18 =

6. 390 + 156,283 + 1,955 =   81 + 17,023 + 950 =

Answers are on page 186.

Lesson 5

Using a Calculator

A pocket calculator is a useful tool to get fast, accurate answers to math problems. If you have a calculator, you can use it to check your work in this book. Below is a drawing of a calculator with descriptions of some of its buttons.

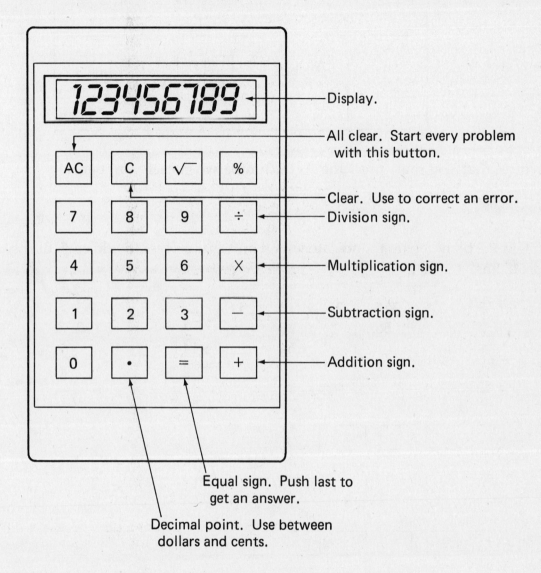

On some calculators the AC button is called C, CE, or CA.

SOLVING ADDITION PROBLEMS ON A CALCULATOR

The following diagram shows how to solve addition problems using a calculator. There are three examples. If you have a calculator, try them, yourself.

Problem	Buttons to Push
25 +36	[AC] [2] [5] [+] [3] [6] [=]
138 + 47	[AC] [1] [3] [8] [+] [4] [7] [=]
6 17 +32	[AC] [6] [+] [1] [7] [+] [3] [2] [=]

Notice that you must press the + button between each number.

EXERCISE

Fill in the blank boxes to indicate which buttons to press to add each of the following problems.

1. 8
   +7

2. 375
   + 66

3. 13
   28
   +64

4. 27
   3
   14
   + 8

5. 2063
   + 49

6. 9
   218
   + 36

1. ☐ ☐ ☐ ☐ ☐

2. ☐ ☐ ☐ ☐ ☐ ☐ ☐ ☐

3. ☐ ☐ ☐ ☐ ☐ ☐ ☐ ☐ ☐ ☐

4. ☐ ☐ ☐ ☐ ☐ ☐ ☐ ☐ ☐ ☐ ☐

5. ☐ ☐ ☐ ☐ ☐ ☐ ☐ ☐

6. ☐ ☐ ☐ ☐ ☐ ☐ ☐ ☐ ☐ ☐

Answers are on page 186.

Lesson 6

Long Column Addition

Exercise

This page gives you a chance to practice your addition skills with longer problems. Remember that you can add the digits in a column in any order. Be sure to use each digit just once. Try to do this page as quickly and accurately as you can.

1.
```
    4      9      7      8      6      3
    6      7      1      5      9      6
    3      1      3      4      3      2
   +8     +5     +4     +4     +7     +2
```

2.
```
    1      8      5      6      4      2
    6      6      2      3      8      4
    8      7      9      2      5      7
    2      3      3      1      1      8
   +5     +7     +2     +7     +6     +9
```

3.
```
   18     26     32     45     73     90
   47     19     17     36     61     57
   36     42     34     63     88     66
  +23    +20    +18    +42    +46    +82
```

4.
```
   61     30     57     81     93     71
   54     82     11     72     18     50
   96     15     93     19     59     48
   70     97     35     66     57     93
  +62    +10    +48    +14    +62    +26
```

5.
```
   12     83     29     44     16     41
   68     24     90     52     79     80
   20     37     12     58     37     12
   37     68     43     16     45     99
  +82    +53    +21    +33    +32    +16
```

Answers are on page 186.

Lesson 7

READING AND WRITING WHOLE NUMBERS

FIRST TEN PLACES AND VALUES

On page 8 you learned the first four places in the whole-number system. Below are the first ten places and values.

Place Names
billions , hundred-millions ten-millions millions , hundred-thousands ten-thousands thousands , hundreds tens units or ones

Place Values
1,000,000,000 , 100,000,000 10,000,000 1,000,000 , 100,000 10,000 1,000 , 100 10 1

COMMAS

Commas (,) separate the digits in whole numbers into groups of three. These commas have no value. They only make numbers easier to read. Each comma in a long number is a reminder of the place name.

In the examples below, the value of the place before each comma is in dark print. Notice how zeros hold places in these examples.

Number	Name
120	one hundred twenty
120,000	one hundred twenty **thousand**
120,000,000	one hundred twenty **million**
3,800,000	three **million**, eight hundred **thousand**
25,460	twenty-five **thousand**, four hundred sixty
36,250,180	thirty-six **million**, two hundred fifty **thousand**, one hundred eighty

Commas can be used with four-digit numbers, but they are not necessary. The number "two thousand five hundred" can be written as 2,500 or 2500. Both ways are correct.

22 Whole Numbers

EXERCISE

Circle the correct answer for questions 1 to 6.

1. Which of the following is two hundred thirty?
   20,013   230   20,030   2,300

2. Which of the following is four thousand, nine hundred?
   4,900   40,009   4,000,900   0049

3. Which of the following is six million, five hundred thousand?
   6,500   6,000,500   6,500,000   65,000,000

4. Which of the following is 5600?
   a. fifty-six
   b. five thousand, six hundred
   c. five million, six hundred
   d. five thousand, six

5. Which of the following is 28,300?
   a. twenty-eight thousand, thirty
   b. two hundred eighty-three thousand
   c. twenty-eight thousand, three hundred
   d. two hundred eighty-three

6. Which of the following is 7,100,000?
   a. seven thousand, one hundred
   b. seven billion, one hundred million
   c. seven million, one hundred
   d. seven million, one hundred thousand

In problems 7 to 12, fill in the blanks to tell the name of each number.

7. 2,300 = two _____, three _____
8. 150,000 = one hundred fifty _____
9. 4,800,000 = four _____, eight hundred _____
10. 7,900 = _____ thousand, _____ hundred
11. 415,000 = four _____ fifteen _____
12. 6,700,000 = _____ million, seven hundred _____

Chapter 1: Addition   23

Problems 13 to 17 have practical examples of large numbers. Fill in the blanks to tell the name of each number.

13. The height, in feet, of the Empire State Building is
    1,250 = one _____, two _____ fifty.
14. The number of feet in one mile is
    5,280 = five _____, two _____ eighty.
15. The official number of unemployed persons in the U.S. in January, 1988, was
    7,000,000 = seven _____.
16. The population of the U.S. in 1986 was
    239,000,000 = two hundred thirty-nine _____.
17. The total cash income for all U.S. farmers in one year was
    $58,000,000,000 = fifty-eight _____ dollars.

In problems 18 to 25, write the whole numbers.

18. eight hundred sixty _____
19. four thousand, one hundred _____
20. nine thousand, six hundred fifty _____
21. sixty-five thousand _____
22. two million _____
23. one hundred eighty thousand _____
24. seven hundred twelve million _____
25. one hundred twenty-five thousand, seven hundred sixty _____

Answers are on page 186.

24 **Whole Numbers**

Lesson 8

ROUNDING OFF

The number 49 is between 40 and 50. 49 is closer to 50 than to 40. 49 **rounded off** to the nearest ten is 50.

The number 213 is between 200 and 300. 213 is closer to 200 than to 300. 213 rounded off to the nearest hundred is 200.

To round off a whole number, follow these steps:

- Underline the digit in the place you want to round off to.
- If the digit to the right of the underlined digit is more than 4, add 1 to the underlined digit. Then write zeros in the places to the right of the digit you underlined.

EXAMPLE 1:   Round off 236 to the nearest ten.

1	2
2<u>3</u>6	240

Step 1.   Underline the digit in the tens place, 3.

Step 2.   Since 6 is the digit to the right of 3 and it is more than 4, add 1 to 3. Write 4 in the tens place and zero in the units place.

- If the digit to the right of the underlined digit is less than 5, leave the underlined digit as it is. Then write zeros in the places to the right of the digit you underlined.

EXAMPLE 2:   Round off 173,926 to the nearest ten thousand.

1	2
1<u>7</u>3,926	170,000

Step 1.   Underline the digit in the ten-thousands place, 7.

Step 2.   Since 3 is the digit to the right of 7 and it is less than 5, leave 7 as it is. Write zeros in the places to the right of 7.

Chapter 1: Addition

EXERCISE

1. Round off each number to the nearest ten.
   536   3,819   14,306   9,052

2. Round off each number to the nearest hundred.
   827   872   26,493   187,628

3. Round off each number to the nearest thousand.
   2,478   18,583   329,627   1,983,455

4. Round off each number to the nearest ten thousand.
   34,927   18,265   74,388   5,276,400

5. Round off each number to the nearest hundred thousand.
   491,000   629,758   3,084,200   18,452,000

6. Round off each number to the nearest million.
   2,750,000   13,429,000   1,978,300   39,563,000

7. In 1985 the population of Mexico City was about 17,300,000. Round off the population to the nearest million. _____

8. Round off the population of Mexico City to the nearest ten million. _____

Answers are on page 186.

Lesson 9

ADDITION WORD PROBLEMS

EXERCISE

In these problems, watch for the key words **sum**, **total**, **more**, **combined**, and **altogether**. These words suggest addition.

After each problem there are questions to help you decide how to solve the problem.

Hint You can use a calculator to check word problems. Be careful with large numbers. There is no comma on a calculator. For example, to enter $15,336 on a calculator, push:

 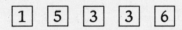

1. A starting office aide working for the city of New York makes $15,336 a year. The highest salary for an office aide is $7,328 more than the starting salary. Find the yearly salary for the highest paid office aide.

   a. What key word suggests addition?
   _____

   b. Solve the problem.

Hint In addition problems with dollars and cents, line up dollars under dollars and cents under cents. On a calculator, be sure to put a decimal point between dollars and cents. For example, to enter the number $1.49 on a calculator, push:

   $\boxed{1}\ \boxed{\cdot}\ \boxed{4}\ \boxed{9}$

2. At a drug store, Danny bought batteries for $1.99 and toothpaste for $1.49. What was the sum of his purchases?

   a. What key word suggests addition?
   _____

   b. Solve the problem.

Hint Some problems give more information than you need.

3. Mr. Gonzales makes $23,750 a year. His wife makes $12,480. Their oldest daughter Sara makes $4,200. Their rent for the year is $2,680. Find the combined yearly income of the three members of the Gonzales family.

   a. What key word suggests addition?
   _____

   b. What number in the problem is **not** someone's income? _____

   c. Solve the problem.

4. Victoria bought a 19-ounce box of cornflakes for $2.89 and a 64-ounce box of detergent for $2.49. What was the total cost of her purchases?
   a. What key word suggests addition?
   _____
   b. For how many items are you asked to find the cost? _____
   c. What numbers in the problem do you **not** need to find the solution? _____
   d. Solve the problem.

Use the information below to answer question 5 and 6.

The list at the right shows the daily costs of caring for a patient at the Riverview Nursing Home.

Housekeeping	$2.09
Food services	5.27
Social services	0.85
Administration	6.29
Nursing	21.06

5. Altogether how much do housekeeping and food services cost for a patient each day at the nursing home?
   a. What key word suggests addition?
   _____
   b. How many numbers do you need to add to solve this problem?
   _____
   c. Solve the problem.

6. Find the total daily costs for a patient at the nursing home.
   a. What key word suggests addition?
   _____
   b. How many numbers do you need to add to solve this problem?
   _____
   c. Solve the problem.

Use the following information to answer questions 7 to 9.

   Jeanne Allen weighs 129 pounds. Her husband Bob weighs 183 pounds. Their son Mike weighs 87 pounds. His cousin Joshua weighs 116 pounds. The Allens' empty car weighs 2250 pounds. For a camping trip, the Allens are taking 235 pounds of equipment, 42 pounds of food supplies, and 36 pounds of clothes.

28 *Whole Numbers*

7. What is the combined weight of Mr. and Mrs. Allen, Mike, and Joshua?
   a. What key word suggests addition?
   _____

   b. For how many people are you asked to add the weight? _____

   c. Solve the problem.

8. Altogether how much do the camping equipment, food supplies, and clothing weigh?
   a. What key word suggests addition?
   _____

   b. For how many items are you asked to add the weight? _____

   c. Solve the problem.

9. Find the total weight of the people, the car, and the equipment, food, and clothing.
   a. What key word suggests addition?
   _____

   b. Solve the problem.

Answers are on page 186.

Lesson 10

BANK CHARGES

EXERCISE

Below is a list of charges from the County Savings Bank. Use this list to answer the questions that follow.

County Savings Bank, Service Charges and Fees	
Checking account, monthly fee	$6.00
Insufficient funds	7.00
Stop payment	9.00
Deposited check returned	1.25
Photocopy of statement	2.00
Savings account under $300.00, monthly fee	1.50

Hint Each month bank customers with checking accounts are charged the $6.00 fee for having a checking account.

1. Fred Guarnieri has a checking account with the County Savings Bank. In April he had to stop payment on a check for damaged merchandise he received. Find the total amount in charges he paid to the County Savings Bank in April.

2. Ellen Johnson also has a checking account with County Savings. One month she lost her statement and asked the bank to send her a photocopy. Also that month, she deposited a check from her brother, but it was returned. Find the total amount in bank charges she had to pay that month.

3. Senior citizens do not have to pay the regular monthly fee for their checking accounts at County Savings. Mr. Berger, who is 78 years old, wrote a check for which he had insufficient funds one month. Find the total amount of his charges that month.

4. Twenty-year-old Grace Gomez has a checking account at County Savings and a savings account with a balance of about $240. Find her total bank charges for a month.

5. Mr. and Mrs. Allen have a checking account with County Savings and a savings account with over $1000 in it. One month they stopped payment on a check they sent to their son because it got lost in the mail. What were their total bank charges for the month?

6. Kevin Graves has a checking account with County Savings and a savings account there with a balance of about $150. He also has a credit card with the bank. In May his credit card interest charges were $23.85. Find the total amount in charges that Kevin paid the bank in May.

Answers are on page 186–187.

Lesson 11

USING A MENU

EXERCISE

Below is a menu from **Carla's Cafe**. Use the menu to answer the questions that follow. Do not worry about tax in these problems.

Sandwiches		Other Items	
Tuna	$1.95	Soup	$1.95
Chicken salad	1.95	Salad	3.50
Grilled cheese	1.25	Soda	0.65
Hamburger	2.65	Coffee or tea	0.50
Cheeseburger	2.95	Pie or cake	1.35

1. Phil had a tuna sandwich and a soda for lunch at Carla's. Find the total amount of his bill.

2. Mrs. McGlynn ordered a grilled cheese sandwich and a cup of tea at Carla's. Find the total amount of her bill.

3. Adrienne and Evelyn had dessert at Carla's. Adrienne ordered apple pie and coffee. Evelyn ordered a piece of chocolate cake and a cup of tea. What was their total bill?

4. Ernie asked for a bowl of soup, a cheeseburger, and a cup of coffee at Carla's. He left a $0.75 tip. How much did he spend altogether including tip?

5. When Alfredo and Jeff went to Carla's, they each ordered a cup of coffee. Alfredo also got a chicken salad sandwich, and Jeff got a piece of pie. What was the total amount of Alfredo's bill?

6. When Paul and Ruth ate at Carla's, Ruth ordered soup, salad, and a soda. Paul ordered soup, a tuna sandwich, a cup of coffee, and a piece of pie. They left a tip of $1.80. How much did they spend at Carla's including tip?

Answers are on page 187.

Lesson 12

BAR GRAPHS

PARTS OF A BAR GRAPH

A **bar graph** has a title and two **scales**. It uses heavy lines, or bars, to compare numbers.

A VERTICAL BAR GRAPH

The bar graph below compares the number of cars sold at Al's Autos in different months. The vertical (up and down) scale at the left tells the number of cars sold. Notice that each guide line represents 10 cars. The horizontal (left to right) scale across the bottom tells the different months. The bars, themselves, run vertically (up and down).

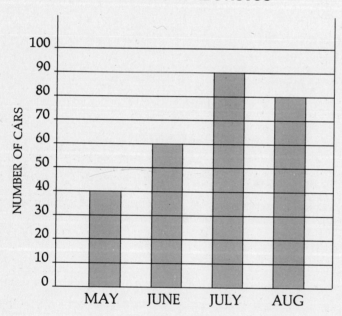

EXERCISE A

Answer the following questions about the graph above.

1. What is the title of the graph? _____
2. The vertical scale is measured in number of _____.
3. What is the lowest number of cars on the vertical scale? _____
4. What is the highest number of cars on the vertical scale? _____
5. How many months are shown on the graph? _____

Study the following example. It will help you with the next problems in this exercise.

34   Whole Numbers

EXAMPLE: How many cars did Al sell in May?

Solution: Find May on the bottom scale. Follow the heavy line above May up to the top. Read directly across on the scale at the left. The heavy line stops at the fourth line up this scale. This line stands for 40. Al sold 40 cars in May.

Use the bar graph on page 34 to answer questions 6 to 12.

6. How many cars did Al sell in June? _____
7. How many cars did Al sell in July? _____
8. How many cars did Al sell in August? _____
9. From the information shown on the graph, in which month did Al sell the most cars? _____
10. In which month did he sell the fewest cars? _____
11. What were the combined car sales for the four months shown on the graph? _____
12. Al sold 35 more cars in September than in August. How many cars did he sell in September? _____

Answers are on page 187.

A HORIZONTAL BAR GRAPH

The bar graph below shows the weights of members of the Larcom family. The bars in this graph run horizontally (across the page). The vertical scale has the name of each family member. The horizontal scale is in pounds. Notice that guide lines run from only 50, 100, and 150, so each represents 50 pounds.

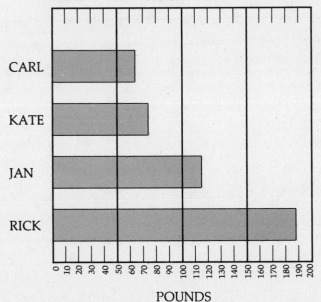

Chapter 1: Addition 35

EXERCISE B

Use the graph on page 35 to answer the following questions.

1. What is the title of the graph? _____
2. The horizontal scale is measured in _____.
3. What is the lowest number of pounds on the horizontal scale? _____
4. What is the highest number of pounds on the horizontal scale? _____
5. The weight of how many people is shown on the graph? _____

The following example will help you with the rest of this exercise.

EXAMPLE: What is the weight of Carl Larcom?

Solution: Find Carl's name on the scale at the left. Follow the heavy line across to the end. Look directly down at the scale at the bottom. The bar ends one fine line to the right of 50. This represents 60. Carl Larcom weighs 60 pounds.

Use the graph on page 35 to answer questions 6 to 11.

6. How much does Kate Larcom weigh? _____
7. How much does Jan Larcom weigh? _____
8. How much does Rick Larcom weigh? _____
9. Find the combined weight of the children, Carl and Kate. _____
10. After six months Jan weighed 18 pounds more than she weighed on the graph. What was her weight after those six months? _____
11. After six months Rick weighed 22 pounds more than he weighed on the graph. What was his weight after those six months? _____

Answers are on page 187.

CHAPTER 2: Subtraction

Lesson 13

SUBTRACTION FACTS

EXERCISE

This exercise gives you practice with the subtraction facts. Write down every answer that you know. Use the guide that follows this exercise to learn the facts you do not already know.

1. 14   9   10   7   5   9   9   11   13   10
   − 7 −8 − 4 −0 −1 −6 −5 − 2 − 8 − 7

2.   6   11   8   7   10   6   12   14   7   6
    −1 − 8 −4 −2 − 2 −6 − 5 − 8 −3 −4

3. 13   11   3   1   15   10   7   9   5   9
   − 9 − 6 −2 −1 − 8 − 5 −6 −2 −3 −9

4. 18   4   17   4   10   13   4   8   13   14
   − 9 −0 − 9 −2 − 1 − 5 −3 −3 − 6 − 5

5. 13   7   9   4   11   16   10   3   15   7
   − 7 −5 −3 −1 − 4 − 7 − 9 −1 − 7 −1

6. 11   9   6   12   17   10   8   12   8   12
   − 3 −0 −5 − 6 − 8 − 3 −7 − 8 −6 − 9

7. 10   11   8   14   7   11   12   5   12   5
   − 6 − 7 −1 − 6 −7 − 9 − 7 −0 − 4 −2

8.   8   13   11   9   16   8   6   15   6   9
    −5 − 4 − 5 −1 − 9 −8 −2 − 9 −3 −7

Chapter 2: Subtraction 37

9. $\quad\begin{array}{r}16\\-8\\\hline\end{array}\quad\begin{array}{r}7\\-4\\\hline\end{array}\quad\begin{array}{r}15\\-6\\\hline\end{array}\quad\begin{array}{r}3\\-3\\\hline\end{array}\quad\begin{array}{r}8\\-2\\\hline\end{array}\quad\begin{array}{r}14\\-9\\\hline\end{array}\quad\begin{array}{r}12\\-3\\\hline\end{array}\quad\begin{array}{r}2\\-1\\\hline\end{array}\quad\begin{array}{r}10\\-8\\\hline\end{array}\quad\begin{array}{r}9\\-4\\\hline\end{array}$

Answers are on page 187.

SUBTRACTION FACTS GUIDE

Try learning a few subtraction facts at a time. Below are the subtraction facts from 18 down to 11.

$\begin{array}{r}18\\-9\\\hline 9\end{array}$

$\begin{array}{r}17\\-9\\\hline 8\end{array}\quad\begin{array}{r}17\\-8\\\hline 9\end{array}$

$\begin{array}{r}16\\-9\\\hline 7\end{array}\quad\begin{array}{r}16\\-8\\\hline 8\end{array}\quad\begin{array}{r}16\\-7\\\hline 9\end{array}$

$\begin{array}{r}15\\-9\\\hline 6\end{array}\quad\begin{array}{r}15\\-8\\\hline 7\end{array}\quad\begin{array}{r}15\\-7\\\hline 8\end{array}\quad\begin{array}{r}15\\-6\\\hline 9\end{array}$

$\begin{array}{r}14\\-9\\\hline 5\end{array}\quad\begin{array}{r}14\\-8\\\hline 6\end{array}\quad\begin{array}{r}14\\-7\\\hline 7\end{array}\quad\begin{array}{r}14\\-6\\\hline 8\end{array}\quad\begin{array}{r}14\\-5\\\hline 9\end{array}$

$\begin{array}{r}13\\-9\\\hline 4\end{array}\quad\begin{array}{r}13\\-8\\\hline 5\end{array}\quad\begin{array}{r}13\\-7\\\hline 6\end{array}\quad\begin{array}{r}13\\-6\\\hline 7\end{array}\quad\begin{array}{r}13\\-5\\\hline 8\end{array}\quad\begin{array}{r}13\\-4\\\hline 9\end{array}$

$\begin{array}{r}12\\-9\\\hline 3\end{array}\quad\begin{array}{r}12\\-8\\\hline 4\end{array}\quad\begin{array}{r}12\\-7\\\hline 5\end{array}\quad\begin{array}{r}12\\-6\\\hline 6\end{array}\quad\begin{array}{r}12\\-5\\\hline 7\end{array}\quad\begin{array}{r}12\\-4\\\hline 8\end{array}\quad\begin{array}{r}12\\-3\\\hline 9\end{array}$

$\begin{array}{r}11\\-9\\\hline 2\end{array}\quad\begin{array}{r}11\\-8\\\hline 3\end{array}\quad\begin{array}{r}11\\-7\\\hline 4\end{array}\quad\begin{array}{r}11\\-6\\\hline 5\end{array}\quad\begin{array}{r}11\\-5\\\hline 6\end{array}\quad\begin{array}{r}11\\-4\\\hline 7\end{array}\quad\begin{array}{r}11\\-3\\\hline 8\end{array}\quad\begin{array}{r}11\\-2\\\hline 9\end{array}$

As you work through this book, look back at this guide until you have memorized every subtraction fact here.

Lesson 14

Subtracting with No Borrowing

SUBTRACTING

The answer to a subtraction problem is called the **difference**. To subtract two numbers, line up the numbers with units under units, tens under tens, and so on. Be sure that the larger number is on top.

Example: 457 − 32 =

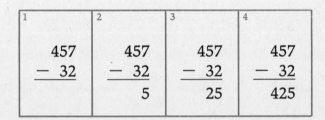

Step 1. Line up the numbers with units under units and tens under tens.
Step 2. Subtract the units. 7 − 2 = 5. Write 5 in the units place.
Step 3. Subtract the tens. 5 − 3 = 2. Write 2 in the tens place.
Step 4. Subtract the hundreds. 4 − nothing = 4. Write 4 in the hundreds place. The answer, or **difference**, is 425.

To solve the last example on a calculator, push:

[AC] [4] [5] [7] [−] [3] [2] [=]

CHECKING

To check a subtraction problem, add the answer to the bottom number in the original problem. The **sum** should be the same as the top number in the original problem.

Subtracting		Checking	
457	top number	32	bottom number
− 32	bottom number	+425	answer
425	answer	457	top number

Chapter 2: Subtraction

EXERCISE

Subtract and check each problem.

1. $427 - 16 =$ $56 - 24 =$ $379 - 127 =$

2. $358 - 6 =$ $165 - 142 =$ $599 - 281 =$

3. $2842 - 742 =$ $2108 - 1106 =$ $3692 - 351 =$

4. $7418 - 205 =$ $6761 - 60 =$ $1435 - 1125 =$

5. $13{,}773 - 2{,}611 =$ $64{,}138 - 12{,}127 =$ $86{,}722 - 2{,}521 =$

6. $42{,}849 - 20{,}735 =$ $98{,}147 - 47{,}043 =$ $28{,}157 - 16{,}042 =$

Answers are on page 187.

Lesson 15

BORROWING

BORROWING ONE TIME

In some subtraction problems, a digit in the bottom number is too big to subtract from the digit in the top number. Then you must **borrow** from the next column in the top number.

You may have learned the terms **renaming** or **regrouping** when you first learned subtraction. Use the terms and methods you are comfortable with. The important thing is getting correct answers.

EXAMPLE: 92 − 75 =

1	2	3

9 2
−7 5
	⁸⁄9 ¹²⁄2
	−7 5
	    7
		⁸⁄9 ¹²⁄2
		−7 5
		  1 7

Step 1. Line up the numbers with units under units and tens under tens.

Step 2. You cannot subtract 5 from 2. Borrow 1 from the 9 in the tens column. This leaves 9 − 1 = 8 in the tens. It gives you ten more units or 10 + 2 = 12 in the units. Subtract the new units. 12 − 5 = 7.

Step 3. Subtract the new tens. 8 − 7 = 1. The answer is 17.

To solve the last example on a calculator, push:

[AC] [9] [2] [−] [7] [5] [=]

EXERCISE A

Subtract and check each problem.

1. 82 − 63 =   47 − 19 =   68 − 59 =   93 − 24 =

2. 46 − 8 =   63 − 24 =   88 − 19 =   34 − 17 =

3. 742 − 36 =   212 − 103 =   361 − 156 =

Answers are on page 187.

BORROWING MORE THAN ONCE

You may have to borrow more than once.

EXAMPLE: 257 − 68 =

```
  1            2              3                4
                              14               14
              4 17           1 4 17          1 4 17
  2 5 7       2 5̸ 7̸         2̸ 5̸ 7̸          2̸ 5̸ 7̸
 −  6 8      −  6 8         −  6 8           −  6 8
                  9            8 9            1 8 9
```

Step 1. Line up the numbers with units under units and tens under tens.

Step 2. You cannot subtract 8 from 7. Borrow 1 from the 5 in the tens column. This leaves 5 − 1 = 4 in the tens. It gives you ten more units or 10 + 7 = 17 in the units column. Subtract the new units. 17 − 8 = 9.

Step 3. You cannot subtract 6 from 4 in the tens column. Borrow 1 from the 2 in the hundreds column. This leaves 2 − 1 = 1 in the hundreds. It gives you ten more tens or 10 + 4 = 14 in the tens column. Subtract the new tens. 14 − 6 = 8.

Step 4. Subtract the new hundreds. 1 − nothing = 1. The answer is 189.

To solve the last example on a calculator, push:
[AC] [2] [5] [7] [−] [6] [8] [=]

Be careful to borrow only when you have to.

EXERCISE B

Subtract and check each problem.

1. 637 − 438 =   443 − 291 =   118 − 79 =

2. 681 − 516 =   375 − 138 =   747 − 258 =

3. 9172 − 560 =   6335 − 827 =   7681 − 599 =

4. 2876 − 1948 =   7509 − 2807 =   5629 − 3497 =

5. 24,582 − 1,991 =   71,556 − 2,497 =   19,233 − 8,315 =

6. 41,629 − 15,462 =   42,386 − 21,523 =   82,716 − 17,947 =

7. 314,214 − 25,064 =   255,612 − 98,751 =   319,634 − 38,536 =

8. 193,212 − 187,446 =   478,446 − 229,381 =   552,683 − 468,294 =

9. 2,481,650 − 1,976,590 =   10,356,000 − 9,548,000 =

Answers are on page 187.

Lesson 16

BORROWING ACROSS ZEROS

ONE ZERO

You cannot borrow from zero. Move left to the next digit that is not a zero. Study the following example carefully.

EXAMPLE: 503 − 28 =

```
   1              2                3                  4
                                    9                  9
                4 10             4 10 13           4 10 13
   5 0 3        5̸ 0̸ 3            5̸ 0̸ 3             5̸ 0̸ 3
 −  2 8       −  2 8           −  2 8            −  2 8
                                                    4 7 5
```

Step 1. Line up the numbers.

Step 2. You cannot subtract 8 from 3, and you cannot borrow from 0. Go on to the hundreds column. Borrow 1 from the 5 in the hundreds column. 5 − 1 = 4. This gives you 10 tens in the tens column.

Step 3. Borrow 1 from the 10 in the tens column. This leaves 10 − 1 = 9 in the tens. It gives you 10 + 3 = 13 in the units.

Step 4. Subtract each column. The answer is 475.

To solve the last example on a calculator, push:

[AC] [5] [0] [3] [−] [2] [8] [=]

EXERCISE A

Subtract and check each problem.

1. 80 − 29 =   60 − 52 =   50 − 18 =

2. 506 − 147 =   308 − 249 =   402 − 337 =

3. 702 − 191 =   205 − 97 =   307 − 286 =

Answers are on page 187.

44   *Whole Numbers*

MORE THAN ONE ZERO

You may have to borrow across more than one zero.

EXAMPLE: 8000 − 264 =

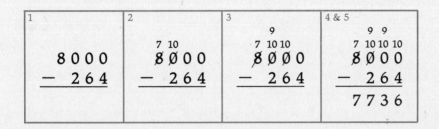

Step 1. Line up the numbers.

Step 2. You cannot subtract 4 from 0, and you cannot borrow from zeros. Borrow 1 from 8 in the thousands column. 8 − 1 = 7. This gives you 10 hundreds in the hundreds column.

Step 3. Borrow 1 from the 10 in the hundreds column. This leaves 10 − 1 = 9 in the hundreds column. It gives you 10 in the tens column.

Step 4. Borrow 1 from the 10 in the tens column. This leaves 10 − 1 = 9 in the tens. It gives you 10 in the units.

Step 5. Subtract each column. The answer is 7736.

To solve the last example on a calculator, push:

| AC | 8 | 0 | 0 | 0 | − | 2 | 6 | 4 | = |

Be sure to borrow only when you have to.

EXERCISE B

Subtract and check each problem.

1. 700 − 432 =   900 − 207 =   400 − 153 =

2. 300 − 88 =   600 − 227 =   500 − 136 =

3. 2500 − 1635 =   1006 − 956 =   3605 − 2846 =

4. 3000 − 1405 =   12,000 − 8,666 =   4000 − 2040 =

5. 10,004 − 9,475 =   12,007 − 10,356 =   40,006 − 18,947 =

6. 38,000 − 19,306 =   50,080 − 8,599 =   90,000 − 24,055 =

7. 407,050 − 267,580 =   308,000 − 247,125 =

8. 2,580,000 − 1,295,000 =   9,000,000 − 856,400 =

9. 6,403,500 − 1,258,700 =   20,540,000 − 16,393,000 =

Answers are on page 188.

Lesson 17

Mixed Subtraction Problems

Exercise

These problems give you a chance to practice the subtraction skills you have learned so far in this book. Subtract and check each problem.

1. 905 − 873 =   936 − 524 =   800 − 256 =

2. 853 − 291 =   400 − 267 =   385 − 249 =

3. 672 − 241 =   905 − 374 =   706 − 298 =

4. 1100 − 235 =   6381 − 590 =   4075 − 238 =

5. 1636 − 493 =   4211 − 508 =   2000 − 483 =

6. 2050 − 1234 =   1651 − 1140 =   7038 − 2527 =

7. 4372 − 2549 =   8127 − 4318 =   8300 − 2650 =

8. 9472 − 1302 =   2055 − 1448 =   6230 − 2115 =

9. 13,225 − 2,340 =   40,563 − 1,942 =   27,045 − 9,928 =

10. 12,000 − 4,507 =   31,942 − 9,756 =   44,500 − 9,386 =

11. 293,428 − 182,516 =   452,000 − 248,118 =

12. 850,436 − 249,998 =   365,227 − 146,483 =

13. 1,000,000 − 396,475 =   6,397,000 − 2,488,000 =

Answers are on page 188.

Chapter 2: Subtraction   47

Lesson 18

SUBTRACTION WORD PROBLEMS

EXERCISE

In the following problems, watch for key phrases like **how much more** and **how much less**. These phrases suggest subtraction. In subtraction problems, remember to put the larger number on top.

1. A starting firefighter in New York makes $25,977 in a year. A starting social worker makes $26,215. How much more does the social worker make in a year?

    a. What key phrase suggests subtraction?
    _____

    b. Solve the problem.

2. Jeanne Allen weighs 129 pounds. Her husband Bob weighs 183 pounds, and their son Mike weighs 87 pounds. How much less does Mrs. Allen weigh than her husband?

    a. What key phrase suggests subtraction?
    _____

    b. What number in the problem do you **not** need to find the answer?
    _____

    c. Solve the problem.

3. A two-door subcompact car rents for $44.95 a week in Florida, for $59.95 a week in Arizona, and for $79.95 a week in Texas. How much more is the weekly rate in Texas than in Florida?

    a. For which two states do you need the costs of renting a car?
    _____

    b. Which number in the problem do you **not** need to find the answer?
    _____

    c. Solve the problem.

Hint Watch for problems with years. You can find the age of someone by subtracting the year the person was born in from the current year. Put the larger number (the most recent year) on top.

4. Clarence Brown was born in 1896. How old was he in 1988?
   a. Which year goes on top?
   _____
   b. Solve the problem.

5. James Naismith invented the game of basketball in 1891. How old was the game in 1988?
   a. Which year goes on top?
   _____
   b. Solve the problem.

Hint Many subtraction problems ask you to compare one thing to another.

6. A 30-inch wide bookshelf valued at $169 was on sale for $129. How much does a buyer save if he gets the shelf on sale?
   a. What number given in the problem do you **not** need to find the answer?
   _____
   b. Solve the problem.

Use the information below to answer questions 7 and 8.

U.S. Population	
1800	5,308,483
1900	75,994,575
1980	226,545,805

7. By how much did the population of the United States grow from 1800 to 1900?
   a. Which population figure do you **not** need for this problem?
   _____
   b. Solve the problem.

Chapter 2: Subtraction 49

8. By how much did the population of the United States grow from 1900 to 1980?

   a. Which population figure do you **not** need for this problem?

   _____

   b. Solve the problem.

9. In February 1987, 66,074 passengers traveled in or out of Burbank Airport. In February 1988, 80,089 passengers used the Burbank Airport. How many more passengers used Burbank Airport in February 1988 than in February 1987?

   a. What key phrase suggests subtraction?

   _____

   b. Solve the problem.

10. In 1985 the number of people 65 and older in the United States was 28,609,000. Experts expect the number of Americans 65 and older to be 34,921,000 in the year 2000. How many more people aged 65 and older will be living in the United States in 2000 than in 1985?

    a. What key phrase suggests subtraction?

    _____

    b. Solve the problem.

Answers are on page 188.

Lesson 19

MIXED ADDITION AND SUBTRACTION WORD PROBLEMS

EXERCISE

In these problems you will use both addition and subtraction. Read carefully. Watch for the words that tell you which operations to use. Following some of the problems are questions to help you decide how to solve the problems.

1. The Howard family's monthly budget is $1180. They spend $275 each month for rent and $430 for food. How much do they have left over each month after paying rent and buying food?

    a. Which monthly expenses are given in the problem? _____

    b. To find the amount they pay for rent and food, do you add or subtract? _____

    c. To find how much is left over, do you add or subtract? _____

    d. Solve the problem.

2. Lois teaches night school. She has 8 women in her class and 11 men. The classroom has 25 student chairs. On a night when all the students are present, how many empty student chairs are there?

3. Tim has to pack the materials in his office because his company is moving to a new location. The box Tim is packing will hold 50 pounds. Tim has already packed 32 pounds of books, 9 pounds of supplies, and 6 pounds of papers. How many more pounds can the box hold?

Chapter 2: Subtraction   51

4. Jim bought a stereo on sale for $329. This is $50 less than the original price. What was the original price?
   a. Was the original price **more** than $329 or **less** than $329? _____
   b. To find the original price, do you add or subtract? _____
   c. Solve the problem.

5. Lynne made $13,600 this year. This is $2,000 less than she made last year. How much did she make last year?

6. A new tire normally selling for $37.95 is on sale for $7.50 off. A set of shock absorbers normally selling for $47.90 is on sale for $10.99 off. Find the total sale price of a tire and a set of shock absorbers.
   a. To find the sale price of a tire, do you add or subtract? _____
   b. To find the sale price of a set of shock absorbers, do you add or subtract? _____
   c. To find the **total** sale price of a tire and a set of shock absorbers, do you add or subtract? _____
   d. Solve the problem.

Use the price list below for problems 7 to 10.

Steve's Auto Service	
Oil change, lube & filter	$14.95
Tire rotation	$18.00
Front-end alignment	$19.50

7. Miriam took her car to Steve's for an oil change, lube, and filter and a front-end alignment. What was the total charge for these services?
   a. How many prices do you need to solve this problem? _____
   b. Solve the problem.

52 **Whole Numbers**

8. The tax on the work on Miriam's car in problem 7 was $2.06. How much change should she get from $40.00?

   a. To find a price including tax, do you add or subtract the tax?
   _____

   b. To find the amount of change, do you add or subtract the price from $40.00?
   _____

   c. Solve the problem.

9. Bob took his car to Steve's for a tire rotation, a front-end alignment, and an oil change, lube, and filter. What was the total cost for these services?

   a. How many prices do you need to solve this problem? _____

   b. Solve the problem.

10. The tax on the work for Bob's car in problem 9 was $3.20. How much change should Bob get from $60.00?

Answers are on page 188.

Lesson 20

Balancing a Checkbook

Below is a page from Mrs. Brown's checkbook register. Each column of the page has a title. Under "Item Number" is the check number or a dash for a deposit or service charge. Under "Date" is the month and day of each transaction. Under "Description" is either the name of the party to whom a check was written or the words **deposit** or **service charge**. Under "Subtractions" are checks and service charges. Under "Additions" are deposits. Under "Balance," each item is either subtracted or added to the amount that was in the account.

		CHECK REGISTER			BALANCE
Item Number	Date	Description	Subtractions	Additions	
					125 13
					− 72 88
62	6/23	County Savings Bank	72 88		52 25
					+ 436 17
—	6/25	deposit		436 17	488 42
					6 00
—	6/28	service charge	6 00		− 482 42

EXERCISE

On July 1 Mrs. Brown wrote check number 63 to Munro Management for $265.00 to pay her rent. On July 7 she wrote check number 64 to General Telephone for $24.79 to pay her phone bill. On July 8 she made a deposit of $436.17. On July 14 she wrote check number 65 to Edison Utilities for $26.54 to pay her electric bill. On July 16 she wrote a check to Mel's Market for $128.76 to pay for groceries. For each item, fill in one line on the checkbook register page above. Find the balance after each transaction.

Answers are on page 189.

Lesson 21

TABLES

A **table** is an orderly set of numbers arranged in rows and columns. Menus and price lists are examples of simple tables.

The table below tells the number of both Democratic and Republican members of the U.S. Congress for two different years. Congress is made up of two "houses." One house consists of representatives. The other house consists of senators.

	1985		1987	
	Democrat	Republican	Democrat	Republican
Representatives	252	182	258	177
Senators	47	53	55	45

Source: Statistical Abstract of the U.S.

EXAMPLE: How many Republican representatives were there in 1985?

Solution: Find 1985. Then look under Republican. The first line below tells the number of representatives. There were 182 Republican representatives in 1985.

EXERCISE

Use the table above to answer the following questions.

1. How many Democratic representatives were there in 1985? _____
2. How many Republican representatives were there in 1987? _____
3. How many Democratic senators were there in 1985? _____
4. How many Republican senators were there in 1987? _____
5. What was the total number of representatives (both Democrat and Republican) in 1985? _____
6. Find the total number of representatives in 1987. _____
7. From 1985 to 1987, the number of Republican representatives dropped by how many? _____
8. For which year shown on the table was the number of Republican senators **less** than the number of Democratic senators? _____
9. In 1985 Republican senators outnumbered Democratic senators by how many? _____
10. In 1987 Democratic representatives outnumbered Republican representatives by how many? _____

Answers are on page 189.

Lesson 22

BAR GRAPHS

You learned about bar graphs on page 34. Notice that the bar graph below shows two different sets of information. It shows both the wins and losses for the Springfield Hawks football team. The wins have a letter W at the top of each bar, and the losses have an L.

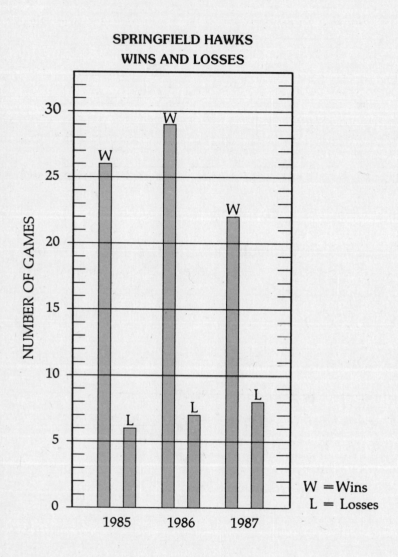

EXERCISE

Use the graph above to answer the following questions.

1. What is the title of the graph?
   _____.

2. The vertical scale tells the number of _____.

3. The letter W stands for _____.

4. The letter L stands for _____.

5. For how many years is information shown? _____

56   *Whole Numbers*

6. How many games did the Hawks win in each of the following years?
   1985 _____   1986 _____   1987 _____

7. How many games did the team lose in each of the following years?
   1985 _____   1986 _____   1987 _____

8. In 1985 the Hawks won how many more games than they lost?
   _____

9. How many more games did the Hawks lose in 1987 than in 1986?
   _____

10. Find the total number of games the Hawks played in each of the following years.
    a. 1985 _____
    b. 1986 _____
    c. 1987 _____

11. In what year shown on the graph did the Hawks win the most games?
    _____

12. In what year shown on the graph did the team lose the least number of games?
    _____

Answers are on page 189.

CHAPTER 3: Multiplication

Lesson 23

MULTIPLICATION FACTS

EXERCISE

This exercise gives you practice with the multiplication facts from 0 to 12. Write every answer that you know. Use the guide that follows this exercise to learn any facts you do not already know.

1. 3   7   10   11   7   0   4   8   2   9
   ×9   ×1   ×6   ×5   ×4   ×3   ×4   ×6   ×9   ×7

2. 7   6   8   12   8   2   10   4   11   5
   ×8   ×4   ×2   ×5   ×9   ×3   ×8   ×2   ×11 ×6

3. 4   6   8   7   12   9   5   1   3   1
   ×11 ×3   ×12 ×11 ×3   ×9   ×7   ×8   ×8   ×0

4. 8   12   4   10   4   3   10   4   12   6
   ×5   ×12 ×3   ×9   ×12 ×11 ×2   ×5   ×7   ×10

5. 9   3   2   7   8   2   10   10   8   3
   ×8   ×5   ×0   ×7   ×4   ×6   ×3   ×11 ×10 ×7

6. 2   6   10   9   8   3   9   3   9   5
   ×5   ×9   ×4   ×12 ×11 ×3   ×2   ×12 ×4   ×11

7. 5   11   2   11   9   1   3   8   0   7
   ×5   ×10 ×7   ×3   ×5   ×5   ×6   ×7   ×4   ×6

8. 7   5   6   9   12   11   9   1   4   3
   ×12 ×0   ×5   ×1   ×6   ×6   ×10 ×1   ×6   ×4

58 Whole Numbers

9.
7	2	11	2	7	5	6	12	11	7
×3	×2	×4	×8	×2	×4	×8	×4	×9	×4

10.
4	6	6	5	9	1	5	7	12	4
×8	×2	×6	×8	×6	×6	×2	×0	×8	×10

11.
5	6	5	6	0	4	3	10	3	1
×12	×11	×9	×7	×6	×9	×2	×10	×9	×7

12.
2	7	8	9	8	7	11	12	5	6
×4	×9	×3	×0	×8	×5	×7	×9	×3	×12

Answers are on page 189.

MULTIPLICATION FACTS GUIDE

Below are the multiplication facts from 1 to 12. Use this table to help learn the facts you do not already know. To find the answer to 4 × 7, first find 4 in the column on the left. Then find 7 in the top row. Follow the row across from 4 and the column down from 7. The row and column meet at the answer, 28.

THE MULTIPLICATION TABLE

	1	2	3	4	5	6	7	8	9	10	11	12
1	1	2	3	4	5	6	7	8	9	10	11	12
2	2	4	6	8	10	12	14	16	18	20	22	24
3	3	6	9	12	15	18	21	24	27	30	33	36
4	4	8	12	16	20	24	28	32	36	40	44	48
5	5	10	15	20	25	30	35	40	45	50	55	60
6	6	12	18	24	30	36	42	48	54	60	66	72
7	7	14	21	28	35	42	49	56	63	70	77	84
8	8	16	24	32	40	48	56	64	72	80	88	96
9	9	18	27	36	45	54	63	72	81	90	99	108
10	10	20	30	40	50	60	70	80	90	100	110	120
11	11	22	33	44	55	66	77	88	99	110	121	132
12	12	24	36	48	60	72	84	96	108	120	132	144

Zeros

The zero multiplication table is easy and important. Zero multiplied by any number is zero. For example, $0 \times 9 = 0$. And, any number multiplied by zero is zero. For example, $4 \times 0 = 0$.

Difficult Multiplication Facts

For most students, the hardest multiplication facts to learn are from 6 to 9. Below are the 16 most difficult multiplication facts. Six of these are repeated in reverse order, for example, both $6 \times 7 = 42$ and $7 \times 6 = 42$ are on this guide.

$6 \times 6 = 36$	$6 \times 7 = 42$	$6 \times 8 = 48$	$6 \times 9 = 54$
$7 \times 6 = 42$	$7 \times 7 = 49$	$7 \times 8 = 56$	$7 \times 9 = 63$
$8 \times 6 = 48$	$8 \times 7 = 56$	$8 \times 8 = 64$	$8 \times 9 = 72$
$9 \times 6 = 54$	$9 \times 7 = 63$	$9 \times 8 = 72$	$9 \times 9 = 81$

As you work through this book, look back at this guide until you have memorized every multiplication fact here.

Lesson 24

MULTIPLYING BY ONE DIGIT WITH NO CARRYING

MULTIPLYING

The answer to a multiplication problem is the **product.** To find the product of a one-digit number and a larger number, first write the one-digit number on the bottom. Then multiply each digit in the top number by the one-digit number. Start with the digit in the units place.

EXAMPLE: 34 × 2 =

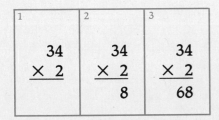

Step 1. Line up the numbers with the one-digit number on the bottom.
Step 2. Multiply 2 by the units digit in the top number. 2 × 4 = 8.
Step 3. Multiply 2 by the tens digit in the top number. 2 × 3 = 6. The product is 68.

CHECKING

To check a product, repeat each step, or use a calculator.

To solve the last example on a calculator, push:
[AC] [3] [4] [×] [2] [=]   or   [AC] [2] [×] [3] [4] [=]

When you use a calculator, either number can come first. However, when you write a multiplication problem, put the smaller number on the bottom.

EXERCISE

Multiply and check each problem.

1. 43 × 3 =   51 × 6 =   60 × 5 =   71 × 8 =

Chapter 3: Multiplication   61

2. 7 × 61 =   2 × 84 =   5 × 91 =   7 × 41 =

3. 81 × 9 =   70 × 6 =   42 × 4 =   32 × 3 =

4. 4 × 62 =   8 × 71 =   5 × 80 =   7 × 91 =

5. 423 × 2 =   511 × 6 =   231 × 3 =   701 × 8 =

6. 7 × 510 =   4 × 621 =   9 × 801 =   2 × 743 =

Answers are on page 190.

Lesson 25

MULTIPLYING BY ONE DIGIT WITH CARRYING

Often the product of two digits is a two-digit number. **Carry** the digit at the left to the next column to the left. Then add the digit you carried to the next product.

EXAMPLE: 79 × 6 =

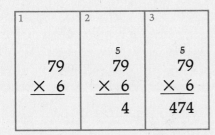

Step 1. Line up the numbers with 6 on the bottom.

Step 2. Multiply 6 by 9. 6 × 9 = 54. Write 4 in the units place and carry 5 to the tens column.

Step 3. Multiply 6 × 7. 6 × 7 = 42. Add the 5 you carried. 42 + 5 = 47. The product is 474.

To solve the last example on a calculator, push:

$\boxed{AC}$ $\boxed{7}$ $\boxed{9}$ $\boxed{\times}$ $\boxed{6}$ $\boxed{=}$ or $\boxed{AC}$ $\boxed{6}$ $\boxed{\times}$ $\boxed{7}$ $\boxed{9}$ $\boxed{=}$

EXERCISE

Multiply and check each problem.

1. 28 × 6 =   73 × 4 =   42 × 8 =   39 × 5 =

2. 3 × 19 =   2 × 50 =   7 × 28 =   6 × 57 =

3. 83 × 5 =   67 × 9 =   75 × 2 =   61 × 8 =

4. 4 × 59 =   2 × 46 =   3 × 18 =   4 × 92 =

5. 114 × 7 =		230 × 9 =		621 × 5 =		542 × 6 =

6. 8 × 791 =		7 × 206 =		3 × 498 =		9 × 566 =

7. 9,208 × 4 =		2,776 × 7 =		6,534 × 6 =		1,892 × 5 =

Answers are on page 190.

Lesson 26

MULTIPLYING BY TWO DIGITS

When you multiply by a two-digit number, first find two **partial products**. Then add them. Remember to start the second partial product under the tens column.

EXAMPLE: 37 × 42 =

1	2	3	4
37 ×42	37 ×42 74	37 ×42 74 148	37 ×42 74 ← partial products 148 1554

Step 1. Line up the numbers. Either 37 or 42 can go on the bottom since they are both two-digit numbers.

Step 2. Multiply 37 by 2. This gives the first partial product, 74.

Step 3. Multiply 37 by 4. This gives the second partial product, 148. Be sure to write the 8 in the tens place because the 4 you are multiplying by represents 4 tens.

Step 4. Add the partial products. The product is 1554.

To solve the last example on a calculator, push:

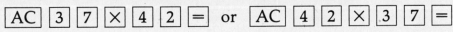

EXERCISE

Multiply and check each problem.

1. 56 × 33 =   17 × 44 =   48 × 66 =

2. 53 × 62 =   84 × 95 =   33 × 74 =

Chapter 3: Multiplication   65

Hint You can use a shortcut if the number you multiply by ends with zero. Put a zero in the partial product under the zero you are multiplying by. Then multiply by the next digit. Below is 92 × 70 multiplied two different ways.

```
       Long method        Shortcut
          92                 92
         ×70                ×70
          00               6440
         644
        6440
```

3. 62 × 20 =   58 × 40 =   77 × 80 =

4. 74 × 14 =   80 × 93 =   91 × 32 =

5. 98 × 27 =   95 × 18 =   56 × 69 =

6. 37 × 16 =   28 × 54 =   31 × 87 =

7. 92 × 24 =   81 × 95 =   73 × 66 =

8. 107 × 71 =   314 × 47 =   892 × 16 =

9. 45 × 218 =   70 × 656 =   63 × 107 =

10. 820 × 38 =   502 × 46 =   930 × 51 =

Whole Numbers

11. 92 × 428 =   73 × 563 =   60 × 227 =

12. 3,520 × 18 =   2,218 × 46 =   1,407 × 33 =

13. 23 × 2,405 =   29 × 8,315 =   82 × 4,630 =

Answers are on page 190.

Lesson 27

MULTIPLYING BY THREE DIGITS

When you multiply by a three-digit number, start the first partial product under the units. Start the second partial product under the tens, and start the third partial product under the hundreds. Then add all three partial products.

EXAMPLE: 1,836 × 214 =

1	2	3	4	5
1,836 × 214	1,836 × 214 7 344	1,836 × 214 7 344 18 36	1,836 × 214 7 344 18 36 367 2	1,836 × 214 7 344 18 36 367 2 392,904

Step 1. Line up the numbers with 214 on the bottom because 214 has fewer digits than 1,836.

Step 2. Multiply 1836 by 4. The partial product is 7344.

Step 3. Multiply 1836 by 1. The partial product is 1836. Write the 6 in the tens column.

Step 4. Multiply 1836 by 2. The partial product is 3672. Write the 2 in the hundreds column.

Step 5. Add the partial products. The product is 392,904.

To solve the last example on a calculator, push:

[AC] [1] [8] [3] [6] [×] [2] [1] [4] [=]

or

[AC] [2] [1] [4] [×] [1] [8] [3] [6] [=]

EXERCISE

Multiply and check each problem.

1.   627 × 519 =   872 × 420 =   386 × 962 =

68   Whole Numbers

2. 497 × 118 =   983 × 603 =   722 × 560 =

3. 732 × 804 =   609 × 513 =   980 × 129 =

4. 359 × 600 =   700 × 705 =   846 × 400 =

5. 3,709 × 218 =   7,143 × 508 =   4,108 × 347 =

6. 941 × 5,072 =   699 × 6,080 =   306 × 7,229 =

Answers are on page 190.

Lesson 28

MULTIPLYING BY 10, 100, AND 1000

Remember that 1 multiplied by any number is that number. It is almost as easy to multiply by 10, 100, or 1000.

To multiply a number by 10, put a zero to the right of the number.

Example: $16 \times 10 = 160$.

To multiply a number by 100, put two zeros to the right of the number.

Example: $435 \times 100 = 43{,}500$.

To multiply a number by 1000, put three zeros to the right of the number.

Example: $99 \times 1000 = 99{,}000$.

EXERCISE

Multiply each problem.

1. $10 \times 37 =$ $100 \times 36 =$ $145 \times 10 =$ $1000 \times 18 =$

2. $100 \times 83 =$ $1000 \times 20 =$ $248 \times 100 =$ $99 \times 10 =$

3. $16 \times 1000 =$ $40 \times 100 =$ $10 \times 250 =$ $100 \times 392 =$

4. $10 \times 1000 =$ $1000 \times 140 =$ $85 \times 100 =$ $200 \times 1000 =$

5. $80 \times 100 =$ $320 \times 10 =$ $410 \times 1000 =$ $10 \times 250 =$

6. A box of electrical parts weighs 72 pounds. Find the weight of 10 boxes. _____

7. One hundred people each gave $35 to a candidate for mayor. How much did they give altogether? _____

8. It costs $2478 a year to educate one child in the Capital City schools. How much does it cost to educate 1000 children? _____

Answers are on page 190.

Lesson 29

MULTIPLICATION WORD PROBLEMS

EXERCISE

You saw three multiplication word problems in the last exercise. In each problem you were given information about one thing, such as the weight of one box. You were asked to find information about several things, such as the weight of ten boxes. This is the most common kind of multiplication word problem.

Watch for similar situations in each of the following problems. In the questions after the problems, you will see the word **per**, which means "for each." These questions should help you find the information about one thing in each problem.

1. Chris can drive 21 miles on one gallon of gasoline. How far can he drive on 13 gallons of gasoline?
    a. How many **miles per gallon** does Chris get with his car? _____
    b. Solve the problem.

Hint Notice the word **total** in the next problem. **Total** was a key word for addition problems. It can also be a key word for multiplication.

2. Bill shipped 19 packages of auto parts. Each package weighed 36 pounds. Find the total weight of the packages.
    a. What was the **weight per package**? _____
    b. Solve the problem.

3. It costs $2486 a year to educate one student in the state of Kentucky. Find the yearly cost of educating a typical class of 30 students in Kentucky.
    a. What is the **cost per student** for one year? _____
    b. Solve the problem.

4. Using the information in problem 3, what is the yearly cost of educating 1000 students in Kentucky? _____

Chapter 3: Multiplication

5. Suzanne earns the minimum wage of $3.35 an hour for a regular 40-hour week. She earns $5.00 an hour for each hour of overtime. How much does she make in a regular 40-hour week?

   a. What is Suzanne's regular **wage per hour**? _____

   b. What number given in this problem do you **not** need? _____

   c. Solve the problem.

6. Guadalupe joined a health club that costs $18 a month. How much does she have to pay for a full year?

   a. What is the club's **cost per month**?
   _____

   b. Solve the problem. (One year has twelve months.)

7. Carl's Carpet Cleaning Service charges $7.95 a room. How much does Carl's company charge to clean the carpets in five rooms?

   a. What is the **cost per room**?
   _____

   b. Solve the problem.

8. Greg can type 67 words in a minute. How many words can he type in 30 minutes?

   a. How many **words per minute** can Greg type? _____

   b. Solve the problem.

Use the following situation to solve problems 9 and 10.

   For a picnic, Elena bought 12 pounds of ground beef for $1.49 a pound. She also bought 10 pounds of spareribs for $1.29 a pound.

9. What was the total cost of the ground beef?

   a. What was the **cost per pound** of the ground beef? _____

   b. How many pounds of ground beef did Elena buy? _____

   c. Solve the problem.

72 **Whole Numbers**

10. What was the total cost of the spareribs?

   a. What was the **cost per pound** of the spareribs? _____

   b. How many pounds of spareribs did Elena buy? _____

   c. Solve the problem.

Answers are on page 190.

Lesson 30

MIXED-OPERATION WORD PROBLEMS

EXERCISE

Each of the following problems requires more than one operation. For example, you may have to multiply and add. Read each problem carefully. Following each problem are questions that will help you decide how to solve the problems.

1. David worked for eight hours at $9 an hour. Then he worked overtime for three hours at $12 an hour. How much did he make altogether?
    a. How much did David make for 8 regular hours? _____
    b. How much did David make for 3 overtime hours? _____
    c. Solve the problem.

2. It costs $59.99 a week to rent a car in Capital City. The daily rate is $24.99. Renting a car for three days at the daily rate is how much more expensive than renting the car for a full week?
    a. How much does it cost to rent a car for three days at the daily rate?
       _____
    b. Solve the problem.

3. Sandy paid $75 as a down payment and $19 a month for twelve months to buy a color television. How much did she pay altogether?
    a. What is the total of Sandy's 12 monthly payments? _____
    b. Should you add or subtract the down payment? _____
    c. Solve the problem.

74 **Whole Numbers**

4. For the performance of a play at the Century Theatre, 525 people paid $6 each for their tickets, and 175 people paid $10 each for their tickets. Find the total value of the ticket sales.

   a. What was the total sale value of all the $6 tickets? _____

   b. What was the total sale value of all the $10 tickets? _____

   c. Solve the problem.

5. Find the total weight of 8 packages that weigh 18 pounds each and 6 packages that weigh 27 pounds each.

   a. What is the total weight of all the 18-pound packages? _____

   b. What is the total weight of all the 27-pound packages? _____

   c. Solve the problem.

6. The Palm Street Block Association hopes to raise $1000 to buy playground equipment. So far, 27 families have each given $25. How much more money does the association need?

   a. Altogether how much did the 27 families give? _____

   b. To find how much more the association needs, do you add or subtract?
   _____

   c. Solve the problem.

7. In a year, the average American eats 83 pounds of chicken, 72 pounds of beef, and 63 pounds of pork. Mr. and Mrs. Chung and their 19-year-old son each eat meat at the national average. Find the total weight of the meat they eat in one year.

   a. In a year, how much meat does each family member eat?
   _____

   b. How many people are in the family?
   _____

   c. Solve the problem.

Answers are on page 190–191.

Chapter 3: Multiplication   75

Lesson 31

THE DISTANCE FORMULA

A formula is an instruction written in the language of mathematics. $D = RT$ is a formula that means, "Distance is equal to rate times time."

D is the distance (usually measured in miles).
R is the rate (usually measured in miles per hour or mph).
T is the time (usually measured in hours).

When two letters in a formula stand next to each other—like R and T—they should be multiplied.

EXAMPLE: Jane drove for 3 hours at a rate of 45 mph. Find the total distance that she drove.

$$D = RT$$
$$D = 45 \times 3 = 135 \text{ miles}$$

Solution: Replace R with 45 and T with 3 in the formula $D = RT$. Jane drove 135 miles.

EXERCISE

Solve each problem.

1. Manny drove for 4 hours at a rate of 63 miles per hour. Find the total distance that he drove.

2. On a camping trip, Mr. Garcia and his daughter walked for 17 hours at a rate of 4 mph. What distance did they walk?

3. On the same trip, Mrs. Garcia and her son walked for 4 hours at a rate of 3 mph. How far did they go?

4. A plane flew for 6 hours at 535 mph. How far did it fly?

5. Find the distance a train travels if it goes for 3 hours at a speed of 79 mph.

6. In town Jack drove for 2 hours at a speed of 11 mph. Then he drove in the country for 3 hours at a speed of 56 mph. How far did he drive altogether?

7. Sam is a truck driver. One day he drove for two hours in a city at a speed of 9 mph. Then he drove for two more hours in the suburbs at a speed of 17 mph. Finally he drove on a highway for four hours at 63 mph. Altogether how far did he drive that day.

Answers are on page 191.

Lesson 32

PERIMETER OF A RECTANGLE

Perimeter is the distance around a flat figure. A **rectangle** is a flat figure with four sides and four square corners. Most doors, windows, and pieces of paper are shaped like rectangles.

To find the perimeter, use the formula $P = 2L + 2W$.

P is the perimeter.
L is the length (usually the long side).
W is the width (usually the short side).

Perimeter is the sum of the four sides. The formula gives a shortcut. It tells you to double the long side, to double the short side, and to add the results together.

Perimeter is measured in units, such as inches, feet, yards, or meters.

EXAMPLE: Find the perimeter of the rectangle shown at the right.

9 feet
5 feet

1.
$P = 2L + 2W$
$P = 2 \times 9 + 2 \times 5$

2.
$P = 18 + 10 = 28$ feet

Step 1. Replace L with 9 and W with 5 in the formula.

Step 2. Multiply 2×9 and 2×5. Then add the results. The perimeter is 28 feet.

EXERCISE

Find the perimeter of each of the following figures.

1. 15 yards, 6 yards

2. 20 meters, 8 meters

78 Whole Numbers

3.
        63 inches

42 inches

4.
        18 feet

9 feet

5. A vegetable garden is 25 feet long and 12 feet wide. How many feet of fencing are needed to go around the garden?

6. A picture is 17 inches long and 11 inches wide. Find the distance around the edge of the picture.

7. A window is 5 feet long and 3 feet high. How many feet of rubber insulation are needed to go around the window?

Answers are on page 191.

Lesson 33

AREA OF A RECTANGLE

Area is a measure of the amount of surface on a flat figure. Finding the amount of carpet needed to cover a floor is an application of area.

To find the area of a rectangle, use the formula $A = LW$.

A is the area.
L is the length.
W is the width.

The formula tells you to multiply the length by the width. Area is always measured in square units, such as square feet or square yards.

EXAMPLE: Find the area of the rectangle shown.

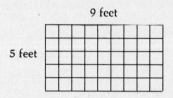

$$A = LW$$
$$A = 9 \times 5 = 45 \text{ square feet}$$

Solution: Replace L with 9 and W with 5 in the formula $A = LW$. Then multiply 9×5. The area is 45 square feet.

Length usually means the longer side, but you can replace L with 5 and W with 9. $5 \times 9 = 45$ gives the same area.

EXERCISE

Find the area of each of the following figures.

1. 18 feet
   10 feet

2. 11 inches
   16 inches

80 Whole Numbers

3.
   14 meters
   7 meters
   [rectangle]

4.
   19 yards
   5 yards
   [rectangle]

5. How many square yards of carpet are needed to cover the floor of a room that is 5 yards long and 4 yards wide?

6. The carpet in problem 5 costs $20 a square yard. What is the total cost of the carpet including a $35 installation fee?

The drawing at the right shows the floor plan of the living room and dining room of the Allen family's new house. Use it to solve problems 7 to 9.

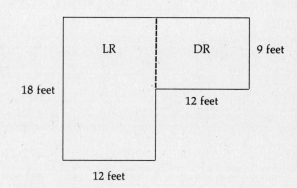

7. What is the area of the living room?

8. What is the area of the dining room?

9. If wood flooring costs $5 a square foot, what is the cost of the flooring required to cover both the living room and the dining room?

Answers are on page 191–192.

Chapter 3: Multiplication 81

Lesson 34

VOLUME OF A RECTANGULAR SOLID

Volume is a measure of the space inside a three-dimensional figure such as a box or a can. Volume is used to measure the amount of water in a pool or the air in a room. A **rectangular solid** is a figure with straight sides and square corners. In addition to length and width, a rectangular solid also has height. A cardboard box is usually in the shape of a rectangular solid.

To find the volume of a rectangular solid, use the formula $V = LWH$.

V is the volume.
L is the length.
W is the width.
H is the height.

Volume is always measured in cubic units, such as cubic inches or cubic meters.

EXAMPLE: Find the volume of the figure pictured at the right.

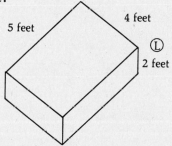

$V = LWH$
$V = 5 \times 4 \times 2 = 40$ cubic feet

Solution: Replace L with 5, W with 4, and H with 2 in the formula. Multiply $5 \times 4 = 20$. Then multiply $20 \times 2 = 40$. The volume is 40 cubic feet.

EXERCISE

Find the volume for each figure.

1.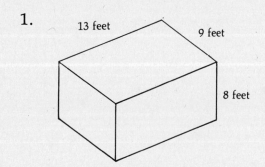

2.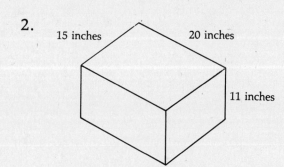

3.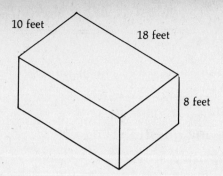

4.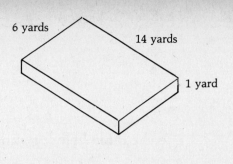

5. The figure at the right shows a pool. Find the volume in cubic feet of water that the pool holds when it is full.

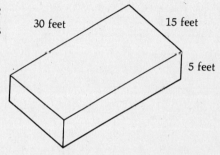

6. The drawing at the right shows an elevator shaft. Find the total volume of the elevator shaft.

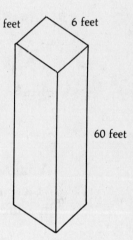

Answers are on page 192.

Lesson 35

TABLES

EXERCISE

Use the table below to answer the following questions.

AVERAGE WEEKLY UNEMPLOYMENT BENEFITS (1986)

State	$	State	$
Alabama	99	Minnesota	166
California	118	New Jersey	156
Florida	123	New York	135
Georgia	109	Texas	160

SOURCE: Statistical Abstract of the U.S.

1. Which state shown in the table paid the highest average weekly unemployment benefits?

2. Which state shown paid the lowest weekly benefits?

3. What was the average amount a worker collected in unemployment benefits for **four** weeks in Alabama?

4. What was the average amount a worker collected in unemployment benefits for four weeks in New Jersey?

5. The average length of time to collect unemployment benefits is 15 weeks. How much did an average worker collect in unemployment benefits for 15 weeks in Texas?

84 Whole Numbers

6. Miguel lives in California. He received the average weekly unemployment benefits for 23 weeks. What total amount did he receive in unemployment benefits?

Use the following information and the table on page 84 to answer questions 7 to 11.

The Jackson family lives in New York. In 1986 they spent $86 a week for food and $278 a month for rent. During four weeks in February the only income for the Jacksons was Mr. Jackson's unemployment benefits.

7. How much did the Jacksons spend for food in four weeks?

8. What total amount did the Jacksons spend for rent and food in four weeks?

9. Mr. Jackson received the average weekly unemployment benefits. How much did he receive in benefits in four weeks?

10. Did the Jacksons get enough in unemployment benefits to cover rent and food expenses for four weeks?

11. If the answer to question 10 was no, how much more did they need? If the answer was yes, how much did they have left over?

Answers are on page 192.

Lesson 36
Bar Graphs

EXERCISE

Use the bar graph below to answer the questions.

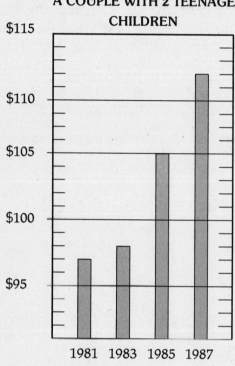

WEEKLY FOOD COSTS FOR A COUPLE WITH 2 TEENAGE CHILDREN

SOURCE: Statistical Abstract of the U.S.

1. What was the average weekly food cost for a couple with two teenage children in
    a. 1981? _____   c. 1985? _____
    b. 1983? _____   d. 1987? _____

2. Find the monthly (four-week) food costs for the family described in the graph in 1983.

3. Find the yearly (52-week) food costs for the family described in the graph in 1987.

86 *Whole Numbers*

4. How much more did food cost for the family described in the graph for one month in 1985 than for one month in 1981?

5. In 1987 the Gibbons family took home $1297 every four weeks. Their food expenses matched those shown on the graph. How much did the Gibbons have left for other expenses besides food each month?

Answers are on page 192.

CHAPTER 4: Division

Lesson 37

DIVISION FACTS

EXERCISE

This exercise gives you practice with the division facts. Write every answer that you know. Use the guide that follows this exercise to learn the facts you do not already know.

1. 4)28   2)14   8)16   2)0   8)48   7)7   3)27   2)6

2. 9)36   1)3   7)42   9)45   4)20   2)4   3)21   3)9

3. 5)10   7)21   9)54   8)0   4)32   7)49   2)18   5)15

4. 8)72   5)40   6)12   3)3   5)20   1)9   5)35   8)64

5. 3)24   3)12   9)63   4)8   1)8   8)40   6)36   7)63

6. 3)15   2)2   5)0   6)48   4)16   4)28   9)27   4)0

7. 6)30   3)6   9)72   2)12   2)10   4)36   2)16   4)24

8. 9)81   7)35   2)8   8)56   9)0   3)18   7)14   5)45

9. 1)5   6)18   8)32   5)25   6)54   6)42   5)5   6)24

10. 8)8   6)0   1)9   5)30   7)56   8)24   1)7   9)18

Answers are on page 192.

DIVISION FACTS GUIDE

For most students, the hardest multiplication facts to learn are the ones with divisors from 6 to 9. Below are the 16 most difficult division facts.

$$6\overline{)36} \quad 6\overline{)42} \quad 6\overline{)48} \quad 6\overline{)54}$$
quotients: 6, 7, 8, 9

$$7\overline{)42} \quad 7\overline{)49} \quad 7\overline{)56} \quad 7\overline{)63}$$
quotients: 6, 7, 8, 9

$$8\overline{)48} \quad 8\overline{)56} \quad 8\overline{)64} \quad 8\overline{)72}$$
quotients: 6, 7, 8, 9

$$9\overline{)54} \quad 9\overline{)63} \quad 9\overline{)72} \quad 9\overline{)81}$$
quotients: 6, 7, 8, 9

As you work through this book, look back at this list until you have memorized every division fact here.

Lesson 38

DIVIDING BY ONE DIGIT

PARTS OF A DIVISION PROBLEM

The answer to a division problem is the **quotient**. A number that divides into another number is the **divisor**. The divisor divides into the **dividend**. We can write these parts of a division problem two different ways:

$$\text{divisor} \rightarrow 5\overline{)35} \begin{array}{l} \leftarrow \text{quotient} \\ \leftarrow \text{dividend} \end{array} \quad \text{or} \quad \begin{array}{c} \text{\textnearrow} \\ \text{dividend} \end{array} 35 \div 5 = 7 \begin{array}{l} \leftarrow \text{quotient} \\ \text{\textnwarrow} \\ \text{divisor} \end{array}$$

DIVIDING

To find a quotient, repeat these four steps until you have finished:

- divide,
- multiply,
- subtract and compare,
- bring down the next number.

In the next example, 6 is the divisor, and 258 is the dividend.

EXAMPLE: $258 \div 6 =$

1 & 2	3 & 4	5 & 6	7
4 $6\overline{)258}$ 24	4 $6\overline{)258}$ $\underline{24}$ 18	43 $6\overline{)258}$ $\underline{24}$ 18 18	43 $6\overline{)258}$ $\underline{24}$ 18 $\underline{18}$ 0

Step 1. **Divide** 6 into 25. $25 \div 6 = 4$ plus a remainder. Write 4 above the 5.

Step 2. **Multiply** 4 by 6. $4 \times 6 = 24$. Write 24 under 25.

Step 3. **Subtract** 24 from 25. $25 - 24 = 1$. And **compare**. The number you get when you subtract must be less than the divisor. 1 is less than the divisor 6.

Step 4. **Bring down the next number**, 8. Write 8 beside 1.

Step 5. **Divide** 18 by 6. 18 ÷ 6 = 3. Write 3 above the 8.

Step 6. **Multiply** 3 by 6. 3 × 6 = 18. Write 18 under 18.

Step 7. **Subtract** 18 from 18. 18 − 18 = 0. And **compare**. 0 is less than the divisor 6. The quotient is 43.

CHECKING

To check a division problem, multiply the quotient by the divisor. The **answer** should be the dividend. For the example:

Dividing	Checking
43	43
6)258	×6
	258

Division is the most complicated of the four basic operations. Be sure you understand each step in the last example before you try the next exercise.

To solve the last example on a calculator, push:

AC 2 5 8 ÷ 6 =

The divisor is always the last number you enter on a calculator.

EXERCISE

Divide and check each problem.

1. 68 ÷ 4 =   364 ÷ 7 =   657 ÷ 9 =   264 ÷ 6 =

2. 280 ÷ 8 =   243 ÷ 3 =   135 ÷ 5 =   138 ÷ 2 =

3. 96 ÷ 3 =   414 ÷ 9 =   182 ÷ 2 =   370 ÷ 5 =

4. 318 ÷ 6 =   100 ÷ 4 =   696 ÷ 8 =   476 ÷ 7 =

5. 434 ÷ 7 =   96 ÷ 2 =   564 ÷ 6 =   180 ÷ 5 =

6. 1248 ÷ 4 =   5944 ÷ 8 =   378 ÷ 3 =   4698 ÷ 9 =

7. 1742 ÷ 2 =   2130 ÷ 5 =   714 ÷ 3 =   6228 ÷ 9 =

8. 4571 ÷ 7 =   5544 ÷ 6 =   1888 ÷ 8 =   576 ÷ 4 =

9. 1645 ÷ 5 =   2838 ÷ 6 =   2436 ÷ 3 =   5808 ÷ 8 =

10. 20,532 ÷ 4 =   38,943 ÷ 9 =   14,568 ÷ 2 =

11. 45,575 ÷ 5 =   17,036 ÷ 2 =   66,258 ÷ 9 =

12. 25,568 ÷ 4 =   35,488 ÷ 8 =   43,470 ÷ 6 =

Answers are on page 192.

Lesson 39

REMAINDERS

Not all division problems come out even. At the end of the last subtraction step, the number you have left is the **remainder**. To check your answer, add the remainder to the product of the quotient and the divisor.

```
       Dividing              Checking
        68   r 3               68
     4)275                     ×4
       24                     272
       ──                     ───
       35                     +3
       32                     ───
       ──                     275
        3
```

EXERCISE

Divide and check each problem. Not every problem has a remainder.

1. 77 ÷ 3 =   369 ÷ 7 =   283 ÷ 6 =   441 ÷ 9 =

2. 93 ÷ 5 =   77 ÷ 2 =   504 ÷ 8 =   287 ÷ 4 =

3. 602 ÷ 7 =   464 ÷ 5 =   290 ÷ 6 =   119 ÷ 2 =

4. 4096 ÷ 8 =   2804 ÷ 3 =   3122 ÷ 9 =   2460 ÷ 4 =

5. 6504 ÷ 9 =   836 ÷ 2 =   1654 ÷ 7 =   2542 ÷ 3 =

6. 6574 ÷ 5 =   11,052 ÷ 4 =   68,183 ÷ 8 =   40,386 ÷ 6 =

Answers are on page 193.

Lesson 40

Zeros As Placeholders

In division problems, every time you bring down a digit, you must put a digit in the quotient. Sometimes you have to hold a place in the quotient with a zero. Look at the zeros in the two examples below. Notice how checking proves that the zeros are necessary.

Example 1:

	Dividing	Checking
	$\begin{array}{r} 40 \\ 8\overline{)320} \\ \underline{32} \\ 00 \\ \underline{0} \\ 0 \end{array}$	$\begin{array}{r} 40 \\ \underline{\times 8} \\ 320 \end{array}$

Example 2:

	Dividing	Checking
	$\begin{array}{r} 208 \\ 3\overline{)624} \\ \underline{6} \\ 02 \\ \underline{0} \\ 24 \\ \underline{24} \\ 0 \end{array}$	$\begin{array}{r} 208 \\ \underline{\times 3} \\ 624 \end{array}$

Exercise

Divide and check each problem. Not every problem requires a zero in the quotient.

1. $210 \div 3 =$ $\quad 247 \div 6 =$ $\quad 278 \div 9 =$ $\quad 363 \div 4 =$

2. $336 \div 8 =$ $\quad 153 \div 5 =$ $\quad 160 \div 2 =$ $\quad 425 \div 7 =$

3. $1224 \div 6 =$ $\quad 1492 \div 4 =$ $\quad 1532 \div 3 =$ $\quad 3672 \div 9 =$

4. $2380 \div 7 =$ $\quad 1012 \div 2 =$ $\quad 2472 \div 8 =$ $\quad 2159 \div 5 =$

5. $1680 \div 8 =$ $\quad 2171 \div 3 =$ $\quad 7263 \div 9 =$ $\quad 4140 \div 6 =$

Answers are on page 193.

Lesson 41

Dividing by Two Digits

To divide by a two-digit number, you have to make guesses. Use the first digit of the divisor to **guess** how many times it will divide into the first two digits of the dividend. Don't worry if your first guess is wrong.

Example: 2368 ÷ 32 =

1 & 2	3 & 4	5 & 6	7
7 32)2368 224	7 32)2368 224 128	74 32)2368 224 128 128	74 32)2368 224 128 128 0

Step 1. **Divide** 236 by 32. Guess how many times 3 goes into 23. 23 ÷ 7 = 3 plus a remainder. Write 7 above the 6.

Step 2. **Multiply** 7 by 32. 7 × 32 = 224. Write 224 under 236.

Step 3. **Subtract** 224 from 236. 236 − 224 = 12. And **compare**. 12 is less than the divisor 32.

Step 4. **Bring down the next number**, 8. Write 8 beside 12.

Step 5. **Divide** 128 by 32. Guess how many times 3 goes into 12. 12 ÷ 3 = 4. Write 4 above the 8.

Step 6. **Multiply** 4 by 32. 4 × 32 = 128. Write 128 under 128.

Step 7. **Subtract** 128 from 128. 128 − 128 = 0. The quotient is 74.

To solve the last problem on a calculator, push:

[AC] [2] [3] [6] [8] [÷] [3] [2] [=]

Exercise

Divide and check each problem.

1. 92 ÷ 23 =   560 ÷ 80 =   108 ÷ 36 =

2. $324 \div 54 =$ $208 \div 98 =$ $380 \div 76 =$

3. $394 \div 52 =$ $243 \div 27 =$ $346 \div 82 =$

4. $192 \div 24 =$ $164 \div 47 =$ $330 \div 55 =$

5. $988 \div 82 =$ $1134 \div 63 =$ $840 \div 40 =$

6. $2714 \div 84 =$ $1742 \div 67 =$ $1530 \div 30 =$

7. $2232 \div 93 =$ $2301 \div 57 =$ $1764 \div 28 =$

8. $2238 \div 77 =$ $7290 \div 90 =$ $3192 \div 57 =$

9. $3350 \div 50 =$ $3312 \div 46 =$ $2648 \div 88 =$

10. $20{,}237 \div 49 =$ $34{,}848 \div 66 =$ $13{,}026 \div 39 =$

11. $7053 \div 23 =$ $28{,}800 \div 60 =$ $8700 \div 75 =$

12. $25{,}756 \div 94 =$ $14{,}510 \div 70 =$ $22{,}092 \div 42 =$

Answers are on page 193.

Lesson 42

DIVIDING BY THREE DIGITS

To divide by a three-digit number, you must also guess. Again, use the first digit of the divisor to guess how many times it will divide into the first one or two digits of the dividend. Do not worry if your first guess is wrong.

EXAMPLE: 4578 ÷ 327 =

1 & 2	3 & 4	5 & 6	7
1 327)4578 327	1 327)4578 $\underline{327}$ 1308	14 327)4578 $\underline{327}$ 1308 1308	14 327)4578 $\underline{327}$ 1308 $\underline{1308}$ 0

Step 1. **Divide.** Guess how many times 3 divides into 4. Write 1 above the 7. You are dividing 327 into 457.

Step 2. **Multiply.** 1 × 327 = 327. Write 327 under 457.

Step 3. **Subtract.** 457 − 327 = 130. And **compare.** 130 is less than the divisor 327.

Step 4. **Bring down the next number.** Write 8 beside 130.

Step 5. **Divide.** Guess how many times 3 goes into 13. Write 4 above the 8.

Step 6. **Multiply.** 4 × 327 = 1308. Write 1308 under 1308.

Step 7. **Subtract.** 1308 − 1308 = 0. The quotient is 14.

To solve the last problem on a calculator, push:

[AC] [4] [5] [7] [8] [÷] [3] [2] [7] [=]

EXERCISE

Divide and check each problem.

1. 384 ÷ 128 =   892 ÷ 216 =   1878 ÷ 313 =

2. 2120 ÷ 424 =   4682 ÷ 641 =   4656 ÷ 582 =

Chapter 4: Division   97

3. 7218 ÷ 802 =   1156 ÷ 578 =   2076 ÷ 439 =

4. 12,720 ÷ 627 =   8896 ÷ 278 =   14,848 ÷ 928 =

5. 18,163 ÷ 443 =   20,805 ÷ 625 =   23,520 ÷ 420 =

Answers are on page 193.

Lesson 43

DIVIDING BY 10, 100, AND 1000

Any number divided by 1 is that number. For example, 8 ÷ 1 = 8 and 27 ÷ 1 = 27. It is almost as easy to divide by 10, 100, or 1000.

To divide a number that ends in one or more zeros by 10, take off one zero at the right.

Example: 150 ÷ 10 = 15.

To divide a number that ends in two or more zeros by 100, take off two zeros at the right.

Example: 48,000 ÷ 100 = 480.

To divide a number that ends in three or more zeros by 1000, take off three zeros at the right.

Example: 326,000 ÷ 1000 = 326.

EXERCISE

Divide each problem.

1. 90 ÷ 10 =   7000 ÷ 1000 =   600 ÷ 100 =

2. 50 ÷ 10 =   4,300 ÷ 100 =   850 ÷ 10 =

3. 12,000 ÷ 1000 =   19,500 ÷ 100 =   230,000 ÷ 1000 =

4. 2,100 ÷ 100 =   500 ÷ 10 =   700,000 ÷ 1000 =

5. 1,650 ÷ 10 =   48,000 ÷ 1000 =   36,500 ÷ 100 =

6. 480 ÷ 10 =   60,000 ÷ 100 =   1,200 ÷ 10 =

7. For ten hours of work, Ellen earns $80. How much does she earn in one hour?

8. The shipping department at Howard's Welding sent 100 welding machines at a total weight of 12,300 pounds. Find the weight of one machine.

9. The budget for 1000 students in Capital City's adult education program is $680,000. What is the cost for one student?

Answers are on page 193.

Lesson

DIVISION WORD PROBLEMS

EXERCISE

You saw three division word problems in the last exercise. In each problem you were given information about several things, such as the weight of several machines. You had to find information about one thing, such as the weight of one machine. This is the most common type of division problem. The key words **share** and **split** also suggest division.

In every division problem, remember to put the amount being divided, the dividend, inside the ⟌ sign. The dividend is the bigger number in whole number problems. Also, the dividend is measured the way the answer is to be measured.

On a calculator, remember to enter the dividend first.

The questions that follow each problem will help you find each dividend.

1. Thirteen office workers shared $6240 in lottery winnings. How much did each of them get?

   a. How is the answer measured, in dollars or number of workers? _____

   b. What is being divided up, $6240 or the 13 office workers? _____

   c. Solve the problem.

2. Mr. Anderson split 222 cubic feet of topsoil among two neighbors and himself. How many cubic feet did each of them receive?

   a. How is the answer measured, in cubic feet or in people? _____

   b. What is being divided up, 222 cubic feet of topsoil or the people?
   _____

   c. How many people will share the soil?
   _____

   d. Solve the problem.

3. Silvia wants to split 92 pounds of berries equally among four baskets. How many pounds of berries will be in each basket?

   a. How is the answer measured, in pounds or baskets? _____

   b. What is being divided up? _____

   c. Solve the problem.

4. Miguel made $315 for 35 hours of work. How much did he make in one hour?

   a. How is the answer measured, in dollars or hours? _____

   b. What is being divided up? _____

   c. Solve the problem.

5. Julie drove 391 miles on 17 gallons of gasoline. She paid $1.09 for each gallon. How many miles did she drive on one gallon of gasoline?

   a. How is the answer measured, in miles, gallons, or dollars? _____

   b. What is being divided up?
   _____

   c. What number in this problem do you **not** need?
   _____

   d. Solve the problem.

6. Mrs. Soto spent ten days in the hospital. The bill was $1800. How much was the bill for one day?

   a. How is the answer measured, in days or dollars? _____

   b. What is being divided up?
   _____

   c. Solve the problem.

7. Frank bought a 120-acre farm for $126,000. What was the price for each acre?

   a. How is the answer measured, in acres or dollars? _____

   b. What is being divided up?
   _____

   c. Solve the problem.

8. Thirty-six cans of soup weigh 432 ounces. Find the weight of one can.
   a. How is the answer measured, in ounces or cans? _____
   b. What is being divided up? _____
   c. Solve the problem.

Answers are on page 193.

Lesson 45

MIXED MULTIPLICATION AND DIVISION WORD PROBLEMS

EXERCISE

The following problems require either multiplication or division. The questions that follow each problem will help you decide what operation to use.

Use the following information for questions 1 and 2.

One weekend the Miller family drove 552 miles. They used 23 gallons of gasoline, and they paid $1.09 for each gallon.

1. How many miles did they drive on one gallon?
   a. How is this answer measured, in miles, gallons, or dollars? _____
   b. What operation do you need, multiplication or division? _____
   c. Solve the problem.

2. How much did they pay for the gasoline?
   a. What is the cost per gallon?
   _____
   b. What operation do you need?
   _____
   c. Solve the problem.

Use the following information for questions 3 and 4.

The Fairfax Theater has a total of 627 seats. Every row in the theater contains 19 seats.

3. How many rows of seats are in the theater?
   a. Should the number of rows be more than 627 or less? _____
   b. Do you need multiplication or division?
   _____
   c. Solve the problem.

104 Whole Numbers

4. A ticket to enter the Fairfax Theater costs $4. When every seat is filled, how much money does the theater collect?
   a. What is the cost per ticket?
   _____
   b. Do you need multiplication or division?
   _____
   c. Solve the problem.

Use the following information for questions 5 to 7.

   Mr. Faust planted 56 acres of his farm with corn. Each acre produced 113 bushels of corn. He was paid $3 for every bushel.

5. Altogether how many bushels of corn did Mr. Faust produce?
   a. How many bushels of corn per acre did the farm produce? _____
   b. Do you need multiplication or division?
   _____
   c. Solve the problem.

6. How much money did he get for all the corn?
   a. What was the price per bushel?
   _____
   b. Do you need multiplication or division?
   _____
   c. Solve the problem.

7. Mr. Faust shared the income from the corn evenly among his three children and himself. How much did each person receive?
   a. What operation does the word **share** suggest? _____
   b. How many people shared the income?
   _____
   c. Solve the problem.

Answers are on page 193–194.

Lesson 46

AVERAGE

Average, or **mean**, is a total divided by the number of items in the total. An average is a value that stands for a set of different numbers, such as a group of test scores or the recent sale prices of a number of houses.

EXAMPLE: There are three children in the Rivera family. Carlos is 20 years old, Carmen is 17, and José is 11. What is their average age?

1	2
20	16
17	3)48
+11	
48	

Step 1. Add the ages of the Rivera children.

Step 2. Divide by the number of children, 3. Their average age is 16.

EXERCISE

Solve each problem.

1. Kevin reached the following scores on Spanish tests: 85, 69, 93, and 77. What was his average score?

2. Nina tutors three students at night school. Kim is 19 years old, Ruth is 38, and Richard is 27. What is their average age?

3. Mrs. Robinson shipped a 9-pound package and a 15-pound package to her son. What was the average weight of the packages?

4. Three houses sold recently on Linden Street. Their prices were $85,000, $72,000, and $89,000. Find the average price.

5. In June Max's phone bill was $14.20. In July it was $39.75. In August it was $28.43. Find Max's average phone bill for these months.

6. At 4:00 P.M., 11 people were eating at Edie's Diner. At 5:00, 20 people were eating, and at 6:00, 26 people were eating. Find the average number of diners each hour.

7. During the first week of December, Reggie's Electronics sold 62 televisions. The second week they sold 85; the third week, 101; and the fourth week, 56. Find the average number of televisions sold per week.

Answers are on page 194.

Lesson 47

UNIT PRICES

EXERCISE

Food is packaged in a variety of amounts. It is difficult to compare prices unless you know the unit price, that is, the cost for one unit, such as an ounce. Following are the prices for several items at Randy's Grocery Store. For each item find the price for one ounce. The first item has been done as an example.

RANDY'S GROCERY STORE

Item	Amount	Price
Pastrami	6 ounces	$2.16
Sausage	8 ounces	$2.32
Bologna	16 ounces	$2.40
Ham	9 ounces	$2.43

Item	Amount	Price
White bread	24 ounces	$1.20
Cheddar cheese	16 ounces	$3.20
Lunch meats	8 ounces	$0.88
Applesauce	33 ounces	$0.99

	ITEM	UNIT PRICE		ITEM	UNIT PRICE
1.	Pastrami	$.36	5.	White bread	_____

$$\begin{array}{r}\$.36\\6\overline{)\$2.16}\end{array}$$

2.	Sausage	_____	6.	Cheddar cheese	_____
3.	Bologna	_____	7.	Lunch meats	_____
4.	Ham	_____	8.	Apple sauce	_____

Answers are on page 194.

Lesson 48

CONVERTING TEMPERATURES

In the United States we measure temperature on the Fahrenheit (F) scale. On this scale, water freezes at 32°, and it boils at 212°. Most other countries use the Celsius, or centigrade (C) scale. On this scale, water freezes at 0°, and it boils at 100°.

CHANGING F° to C°

To change a Fahrenheit temperature to the corresponding Celsius temperature, follow these steps:

- subtract 32
- multiply by 5
- divide by 9

EXAMPLE: **Change 77° Fahrenheit to Celsius.**

1	2	3
77° −32° ――― 45°	45° ×5 ――― 225°	25° 9)225° 18 ―― 45 45 ―― 0

Step 1. Subtract 32 from 77.
Step 2. Multiply 45 by 5.
Step 3. Divide 225 by 9. 77°F = 25°C.

EXERCISE A

Change each of the following Fahrenheit temperatures to Celsius.

1. 50°F =   95°F =   68°F =   59°F =   104°F =

2. 41°F =   86°F =   140°F =   122°F =   212°F =

Answers are on page 195.

CHANGING C° TO F°

To change a Celsius temperature to the corresponding Fahrenheit temperature, follow these steps:

- multiply by 9
- divide by 5
- add 32

EXAMPLE: Change 15° Celsius to Fahrenheit.

1	2	3
15 ×9 --- 135	27 5)135 10 --- 35 35 --- 0	27 +32 --- 59

Step 1. Multiply 15 by 9.
Step 2. Divide 135 by 5.
Step 3. Add 27 and 32. 15°C = 59°F.

EXERCISE B

Change each of the following Celsius temperatures to Fahrenheit.

1. 200°C =   80°C =   30°C =   75°C =   5°C =

2. 95°C =   60°C =   100°C =   150°C =   10°C =

Answers are on page 195.

110 Whole Numbers

Whole Numbers Review

These problems will help you decide what you need to review about whole numbers.

1. 3,618 + 2,371 =
2. 479 + 936 =
3. 588 + 2,942 + 4,055 =
4. 516,279 + 38,668 =
5. 437 − 219 =
6. 8,246 − 5,178 =
7. 20,400 − 12,374 =
8. 310,000 − 197,488 =
9. 426 × 8 =
10. 27 × 683 =
11. 6 × 28,049 =
12. 436 × 287 =
13. 4257 ÷ 9 =
14. 4928 ÷ 16 =
15. 1230 ÷ 82 =
16. 3087 ÷ 49 =

17. Write the number four hundred seventy. _____
18. Write the number seventy-one thousand. _____
19. Write the number three hundred six thousand. _____
20. Write the number one million, eight hundred fourteen thousand. _____

21. Round off 186 to the nearest ten. _____
22. Round off 2,427 to the nearest hundred. _____
23. Round off 11,378 to the nearest thousand. _____
24. Round off 126,300 to the nearest ten thousand. _____

25. Jeanne bought a 5-ounce tube of toothpaste for $1.59 and an 11-ounce bottle of shampoo for $3.29. What was the total cost of her purchases?

26. Bill Thomas takes home $472 a week. His wife Reba takes home $468. Their son Mike takes home $234 a week. The monthly mortgage payments on their house is $365. What is the combined weekly income of Mr. and Mrs. Thomas and their son?

27. In 1986 the population in the urban area including Dallas and Ft. Worth was 3,655,000. The population in the area including Houston and Galveston was 3,634,000. Altogether what was the population of these two areas?

28. In 1987 John earned $19,270. In 1988 he made $1,560 more. How much did John make in 1988?

29. David Rivera wants to buy a videocassette recorder. The model he wants costs $229.97 at Steve's Stereo. The same model costs $264.98 at Discount City. How much more does the recorder cost at Discount City?

30. The United States became an independent country in 1776. For how many years had the United States been independent by 1988?

31. Eric bought a gallon of exterior house paint for $6.99 and a gallon of deck paint for $8.99. Altogether the tax for the paint was $.72. How much change did Eric get from $20.00?

Use the table below to answer questions 32 and 33.

TYPICAL WEEKLY EARNINGS OF A U.S. FAMILY (IN DOLLARS)

	1983	1984	1985	1986
Married couple with one worker	350	371	385	393
Married couple with two workers	615	650	684	712

SOURCE: Statistical Abstract of the U.S.

32. By how much did the weekly earnings of a typical married couple with one worker increase from 1983 to 1986?

33. In 1985 how much more were the weekly earnings of a typical family with two workers than of a family with one worker?

Use the bar graph below to answer questions 34 and 35.

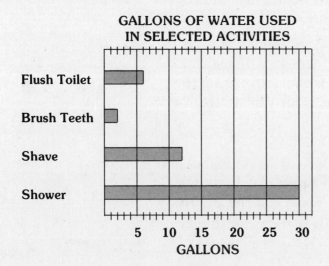

34. In the morning Manny flushes the toilet, brushes his teeth, showers, and shaves. How many gallons of water does he use?

35. Washing dishes by hand takes about 20 gallons of water. A shower requires how much more water than washing dishes?

36. The Millers sold their house for $68,500. This is $9,000 more than they paid for it. How much did they pay for their house?

37. Carla paid $15.95 a month for 24 months to buy dining room furniture. How much did she pay altogether?

Chapter 4: Division   113

38. Find the total weight of 30 cartons if each carton weighs 17 pounds.

39. Max bought four new tires at a cost of $32.50 each. The tax on the tires was $8.45. How much change did he get from $150.00?

40. Nina drove for 3 hours at an average speed of 59 mph. Approximately how far did she drive?

41. The formula for finding the perimeter of a rectangle is $P = 2L + 2W$, where P is the perimeter, L is the length, and W is the width. Find the perimeter of a living-room floor that is 18 feet long and 13 feet wide.

42. The formula for finding the area of a rectangle is $A = LW$, where A is the area, L is the length, and W is the width. What is the area of the living-room floor in problem 41?

43. Paul paid $1200 down and then $45 a month for 36 months for a used car. How much did he pay altogether?

44. Millie paid $15.90 for 6 pounds of fish. What was the price for one pound of fish?

45. For their parents' anniversary, the three Johnson children shared the cost of sending their parents to a lake-side cabin for a week. The cost of the trip was $957. How much did each of the children pay?

Use the following information for problems 46 to 48.

The Hung family drove a total of 494 miles one weekend. During the trip they bought 26 gallons of gasoline for $27.82 and used it all.

46. What was their gasoline mileage? (How far did they drive on one gallon of gasoline?)

47. How much did one gallon of gasoline cost?

48. Altogether the Hungs drove for 13 hours. What was their average speed in miles per hour?

49. Celia works part time. Friday night she worked 6 hours. Saturday she worked 10 hours, and Sunday she worked 5 hours. What average number of hours did she work per day?

50. Plastic Products, Inc., shipped 100 identical boxes of parts at a total weight of 1800 pounds. How much did each box weigh?

Check your answers on page 195–196.

WHOLE NUMBERS REVIEW RECORD

Write the number of problems you answered correctly: _____

Write the number of problems you answered incorrectly: _____

Use the Lesson Guide that follows to find out which lessons you should review, if any. This Lesson Guide can help you make a plan for reviewing whole numbers. For each problem in the Whole Numbers Review, the lesson that teaches about that kind of problem is listed.

WHOLE NUMBERS REVIEW LESSON GUIDE

Problem Number	1	2	3	4	5	6	7	8	9	10	11	12
Lesson Number	3	4	4	4	15	15	16	16	25	26	25	27
Problem Number	13	14	15	16	17	18	19	20	21	22	23	24
Lesson Number	38	41	41	41	7	7	7	7	8	8	8	8
Problem Number	25	26	27	28	29	30	31	32	33	34	35	36
Lesson Number	9	9	9	9	18	18	19	21	21	22	22	19
Problem Number	37	38	39	40	41	42	43	44	45	46	47	48
Lesson Number	29	29	30	31	32	33	30	44	44	45	45	45
Problem Number	49	50										
Lesson Number	46	44										

PART B

DECIMALS

Decimals Pretest

Chapter 1   Addition and Subtraction
Chapter 2   Multiplication
Chapter 3   Division

Decimals Review

Decimals Pretest

These problems will help you decide what you need to study most about decimals. Solve every problem that you can. Then check your answers. Finally, fill in the Decimals Pretest Record on the next page. It will show you which pages you should study.

Addition and Subtraction

1. $.7 + .9 =$
2. $.36 + .419 =$

3. $1.2 + .488 + .57 =$
4. $4.3 + 1.52 + 6 =$

5. $.06 + .0075 + .0198 =$
6. $25.3 + 2.77 + 1.896 =$

7. $.9 - .6 =$
8. $.8 - .26 =$

9. $4 - 1.37 =$
10. $2 - .095 =$

11. $.685 - .59 =$
12. $23.2 - 16.478 =$

Multiplication

13. $.8 \times .9 =$
14. $.6 \times .5 =$

15. $.04 \times .2 =$
16. $1.36 \times 7 =$

17. $21.4 \times .9 =$
18. $36.2 \times .15 =$

19. $1.8 \times .036 =$
20. $.09 \times .014 =$

21. $2.6 \times 100 =$
22. $6.2 \times 1000 =$

Division

23. 10.4 ÷ 8 =

24. 2.88 ÷ 12 =

25. 2.24 ÷ .7 =

26. .0168 ÷ .06 =

27. 28.8 ÷ 3.2 =

28. 12.6 ÷ .18 =

29. 60 ÷ 1.5 =

30. 52 ÷ .04 =

31. 26.2 ÷ 100 =

32. 4.8 ÷ 1000 =

Check your answers on page 197.

DECIMALS PRETEST RECORD

Section	Circle the numbers of the problems you got right	If your score is	Study pages
Addition and Subtraction	1 2 3 4 5 6 7 8 9 10 11 12	0 - 9 10 - 12	120 - 142 126 - 129 134 - 142
Multiplication	13 14 15 16 17 18 19 20 21 22	0 - 7 8 - 10	143 - 159 148 - 159
Division	23 24 25 26 27 28 29 30 31 32	0 - 7 8 - 10	160 - 179 171 - 179

For each problem in the Decimals Pretest, the lesson that teaches about that kind of problem is listed in this guide.

DECIMALS PRETEST LESSON GUIDE

Problem Number	1	2	3	4	5	6	7	8	9	10	11	12
Lesson Number	2	2	2	2	2	2	5	5	5	5	5	5
Problem Number	13	14	15	16	17	18	19	20	21	22	23	24
Lesson Number	12	12	12	12	12	12	12	12	13	13	21	21
Problem Number	25	26	27	28	29	30	31	32				
Lesson Number	22	22	22	22	23	23	26	26				

CHAPTER 1: Addition and Subtraction

Lesson 1

INTRODUCING DECIMALS

Decimals are special kinds of fractions. They tell parts of whole things. A decimal divides a whole into ten parts, or one hundred parts, or one thousand parts, and so on.

You have used decimals already in this book. Our money system uses decimals. A dime is one of the ten equal parts in a dollar. A penny is one of the one hundred equal parts in a dollar.

DECIMAL PLACES

Decimal places are at the right of the decimal point. The chart below shows the first six whole-number places and the first six decimal places.

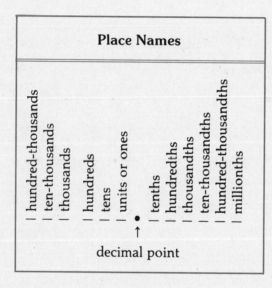
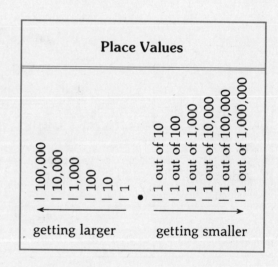

DECIMAL PLACE VALUE

As you move to the right in the decimal system, each place means that the whole has been divided into more parts. The value of decimal places gets smaller as you go to the right.

Think about the number .75 and the value of each digit in it.

- The digit 7 is in the tenths place. This means that one whole was divided into 10 equal parts. 7 has a value of 7 of the 10 equal parts.
- The digit 5 is in the hundredths place. This means that one whole was divided into 100 equal parts. 5 has a value of 5 of the 100 equal parts.

Now, think about $.75.

- The digit 7 has a value of 7 of the 10 dimes that make up a dollar. The value of the 7 is 7 × 10c = 70c.
- The digit 5 has a value of 5 of the 100 pennies that make up a dollar. The value of the 5 is 5 × 1c = 5c.

SIZES OF DECIMALS

Following are lines that show the size of decimals. Each line is measured in inches or parts of an inch.

Line A, 4 inches: **Line A** ├────┼────┼────┼────┤
4 is a whole number.

Line B, .4 inch: **Line B** ├┼┼┼┤││││││
    Line B is a .4 inch, or four tenths inch, long. If we divide an inch into 10 equal parts, .4 inch is 4 of those 10 parts.

Line C, .04 inch: **Line C** ├│││││││││││
    The very short line C is .04, or four hundredths, inch long. If we divide an inch into 100 equal parts, .04 is 4 of those 100 parts.

Imaginary line D, .004 inch. Line D is too short to show. If we divide an inch into 1000 equal parts, .004 is 4 of those 1000 parts.

Imaginary line E, .0004 inch. Line E is **much** too short to show. If we divide an inch into 10,000 equal parts, .0004 is 4 of those 10,000 parts.

COUNTING DECIMAL PLACES

To count decimal places, count the digits to the right of the decimal point. The number .0026 has four decimal places, since all four digits are to the right of the point.
    The number 23.5 has only one decimal place, since only the digit 5 is to the right of the point. 23.5 is called a **mixed decimal**. A mixed decimal has a whole number and a decimal fraction.

EXERCISE

1. Circle the numbers that have one decimal place, or tenths.
   10.3   1.03   .4   7.54   7.4   .8

2. Circle the numbers that have two decimal places, or hundredths.
   2.356   23.56   .75   1.017   1.17   17.1

Chapter 1: Addition and Subtraction

3. Circle the numbers that have three decimal places, or thousandths.
   2.9921   6.053   403.5   .0488   .488   1.675

4. Circle the digit in the tenths place in each number.
   4.5   .6   20.15   2.678   .35   2.8

5. Circle the digit in the hundredths place in each number.
   .05   .24   2.16   .3875   2.87   .367

6. Circle the digit in the thousandths place in each number.
   .125   2.6087   .0093   1.003   .749   25.128

Use the number **3.457** to answer questions 7 to 10. Fill in the blanks.

7. 3 is in the _____ place. 3 has a value of _____.
8. 4 is in the _____ place.
   4 has a value of 4 out of _____ equal parts in one whole.
9. 5 is in the _____ place. 5 has a value of 5 out of
   _____ equal parts in one whole.
10. 7 is in the _____ place. 7 has a value of 7 out of
    _____ equal parts in one whole.

Answers are on page 197.

Lesson 2

Adding Decimals

LINING UP DECIMAL POINTS

Adding decimals is like adding whole numbers. To add decimals, line up the decimal points. This puts units under units, tenths under tenths, hundredths under hundredths, and so on.

Example 1: .9 + .5 =

```
1                    2
     .9 tenths            units
                          .9 tenths
    .9                    .9
   +.5                   +.5
                         1.4
```

Step 1. Line up the decimal points.

Step 2. Add the tenths. When you carry, the 1 goes over to the units column. The sum is 1.4.

To solve the last example on a calculator, push:

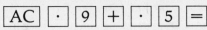

Notice that the calculator automatically puts a zero before the decimal point.

DECIMAL POINTS AND PERIODS

The end of the last sentence in Step 2 above has something that could be confusing: 1.4. This number does not have two decimal points. Only the point between 1 and 4 is a decimal point. The last dot is a period at the end of the sentence.

WHOLE NUMBERS AND DECIMAL POINTS

Remember that the decimal point comes at the right of a whole number. When you add a column of numbers, put a decimal point at the right of every whole number even if it does not appear in the problem.

EXAMPLE 2:   3.62 + .756 + 8 =

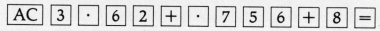

Step 1. Line up the numbers, and put a decimal point at the right of the whole number 8.

Step 2. Add each column. The sum is 12.376.

To solve the last example on a calculator, push:

`AC` `3` `.` `6` `2` `+` `.` `7` `5` `6` `+` `8` `=`

Again, the calculator does some things for you. It puts the 0 before .756, and it puts the decimal point after 8.

EXERCISE

For problems 1 to 5, choose the solution that is set up correctly. Then add the correct solution.

1. 2.5 + .36 =
   - (a) 2.5
            +.3 6
   - (b) 2.5
            +.36
   - (c) 2.5
            +.36

2. .8 + .43 + .128 =
   - (a) .8
            .43
            +.128
   - (b) .8
            .43
            +.128
   - (c) .8
            .43
            +.128

3. 9 + 2.37 =
   - (a) 9.
            +2.37
   - (b) 9.
            +2.37
   - (c) 9.
            +2.37

4. .03 + .4 + .069 =
   - (a) .03
            .4
            +.069
   - (b) .03
            .4
            +.069
   - (c) .03
            .4
            +.069

124 *Decimals*

5. 1.3 + .88 + 4 =   (a) 1.3   (b)   1.3   (c)   1.3
                           .88   .88   .8 8
                         + 4.   +4.   + 4.
                         ─────   ────   ─────

Solve each problem.

6.   .4 + .7 =   .36 + .9 =   .385 + .76 =

7.   6.43 + .459 =   7.87 + 85.4 =   396.4 + 27.5 =

8.   30.1 + 6.04 =   16.7 + 3.359 =   .486 + .9073 =

9.   2.75 + 14 =   9 + 1.886 =   1.422 + 34.32 =

10. 5.778 + 16.43 + .276 =   829.3 + 14.78 + 9.556 =

11. 6.15 + .385 + 9 =   .485 + .13 + .7 =

12. 4.054 + 7.69 + 1.783 =   2 + .48 + .992 =

Answers are on page 197.

Chapter 1: Addition and Subtraction

Lesson 3

READING AND WRITING DECIMALS

DECIMAL NAMES

Decimals get their names from the **last** decimal place to the right. Look at these examples. Notice the way zeros are used to give some of the decimals the correct number of places.

Decimal	Name	Remarks
.4	four **tenths**	only one decimal place
.07	seven **hundredths**	two decimal places
.30	thirty **hundredths**	two decimal places
.009	nine **thousandths**	three decimal places
.012	twelve **thousandths**	three decimal places
.0011	eleven **ten-thousandths**	four decimal places

EXERCISE A

1. Which of the following decimals is two tenths?
   210   .2   2.10   .02

2. Which of the following decimals is eight hundredths?
   .08   .80   8100   8.100

3. Which of the following decimals is fifteen thousandths?
   15,000   15.000   .015   .150

4. Which of the following decimals is twenty-seven hundredths?
   .027   .27   27.100   2700

5. Which of the following decimals is two hundred nineteen thousandths?
   219,000   .20019   219,000   .219

6. Which of the following decimals is twenty-six ten-thousandths?
   .2610   260,000   .0026   26.10

Write in each blank the word that completes the name of the decimal.

7. .5 = five _____   .6 = _____ tenths
8. .09 = nine _____   .13 = _____ hundredths
9. .12 = twelve _____   .004 = _____ thousandths
10. .025 = twenty-five _____   .019 = _____ thousandths

Answers are on page 197.

MIXED DECIMALS

The number 26.048 includes a whole number and a decimal. The whole number is 26, and the decimal is .048. The number 26.048 is called a **mixed decimal**. Read 26.048 as "twenty-six **and** forty-eight thousandths." Notice that the word **and** separates the whole number from the decimal.

Hint Remember that the decimal point is not a place. It only separates the whole number places from the decimal places.

Exercise B

1. Which of the following mixed decimals is four and nine tenths?
   4910   4.910   4.9   49.10

2. Which of the following mixed decimals is eighty and two hundredths?
   80.02   80.200   .82   82.100

3. Which of the following mixed decimals is three and fifteen thousandths?
   315,000   3.15000   .315   3.015

Fill in the blanks to complete the names of the mixed decimals below.

4. 4.1 = four and one _____
5. 8.6 = _____ and _____ tenths
6. 2.16 = two and sixteen _____
7. 20.19 = _____ and _____ hundredths
8. 17.003 = seventeen and _____ thousandths
9. 5.014 = _____ and _____ thousandths

Problems 10 to 12 have practical examples of decimals. Fill in the blanks to complete the names.

10. 2.2 = _____ and _____ tenths is the number of pounds in a kilogram.
11. 2.54 = two and fifty-four _____ is the number of centimeters in one inch.
12. 1.6 = _____ and _____ is the number of kilometers in one mile.

Write each of the following as decimals.

13. three tenths = _____
14. thirteen hundredths = _____
15. four thousandths = _____
16. five ten-thousandths = _____
17. sixteen hundredths = _____

Write each of the following as mixed decimals.

18. twelve and six tenths = _____
19. five and nine thousandths = _____
20. sixty and eight tenths = _____
21. three and eleven thousandths = _____
22. fourteen and six hundredths = _____

Answers are on page 197.

Lesson 4

COMPARING DECIMALS

You know that $.50 is larger than $.45. It is more difficult to compare .5 and .45. To compare decimals, first give the decimals you are comparing the same number of places. You can add zeros to the right of a decimal without changing the value. .6 is the same as .60.

Six dimes have the same value as 60 pennies. In each number, the digit 6 is in the tenths place.

EXAMPLE: Which number is larger, .5 or .45?

1	2
.5 = .50	.50 is larger than .45, so .5 is the larger number.

Step 1. .45 has two decimal places. Put a zero to the right of .5 to give it two places.

Step 2. Fifty hundredths is bigger than forty-five hundredths. .5 is the larger number.

EXERCISE

In problems 1 to 3 circle the larger number in each pair.

1. .8 or .83   .6 or .49   2.5 or 2.55
2. 3.4 or 3.09   .038 or .04   .047 or .03
3. 10.2 or 8.99   .106 or .02   .106 or .2

In problems 4 to 6 circle the largest number in each group.

4. .7, .07, or .717   .3, .05, or .035
5. .106, .11, or .016   1.02, 1.202, or 1.2
6. 3.81, 3.188, or 3.88   .04, .413, or .41

7. Which is heavier, a package that weighs .12 pound or a package that weighs .095 pound? _____

8. Which is longer, a piece of pipe that is 1.06 meters long or a piece of pipe that is 1.1 meters long? _____

Answers are on page 197.

Lesson 5

SUBTRACTING DECIMALS

To subtract decimals, first line up the decimal points. This puts tenths under tenths, hundredths under hundredths, and so on. Be sure to put the larger number on top. Add zeros to give the numbers the same number of decimal places. Then subtract.

EXAMPLE 1:   .72 − .3 =

1	2	3
.72 −.3	.72 −.30	.72 −.30 .42

Step 1.   Line up the decimal points.

Step 2.   Put a zero at the right of .3. This gives both numbers two decimal places. It does **not** change the value of .3.

Step 3.   Subtract. The difference is .42.

To solve the last example on a calculator, push:

AC . 7 2 − . 3 =

EXAMPLE 2:   6 − 2.87 =

1	2	3
6. −2.87	6.00 −2.87	9 5 10 10 6̶.0̶ 0̶ −2.8 7 3.1 3

Step 1.   Put a decimal point at the right of 6. Then line up the decimal points.

Step 2.   Put two zeros at the right of 6. This gives both numbers two decimal places.

Step 3.   Borrow and subtract. You can borrow across the decimal point. The difference is 3.13.

To solve the last example on a calculator, push:

AC 6 − 2 . 8 7 =

130   *Decimals*

EXERCISE

Subtract and check each problem.

1. .3 − .21 = .24 − .015 = .74 − .3 =

2. 2.7 − 1.38 = 14.2 − 6.135 = .8 − .45 =

3. .37 − .288 = .6 − .093 = 12.2 − 1.77 =

4. 4 − 1.276 = 8.66 − 3 = 27.2 − 25.48 =

5. 2 − 1.496 = 5 − .38 = .73 − .6 =

6. 21 − 19.4 = 8.2 − 8.119 = 28.6 − 13 =

7. .06 − .035 = 18 − 2.344 = 1.288 − .7 =

Answers are on page 198.

Lesson 6

MIXED ADDITION AND SUBTRACTION

EXERCISE

This exercise gives practice with both adding and subtracting decimals. Remember to put a decimal point at the right of each whole number. Remember also to use zeros in decimal subtraction problems to give both numbers the same number of places.

Solve each problem.

1. $5.5 + 4.95 =$ $5.5 - 4.95 =$

2. $10 - 0.66 =$ $.81 - .793 =$

3. $6.4 + .64 + 64 =$ $15 - 2.694 =$

4. $3.8 + 1.78 =$ $.01 - .0037 =$

5. $1,200 - 846.32 =$ $.8 + .856 + .07 =$

6. $1.3 + 1.265 =$ $43.9 + .439 =$

7. $6.2 + .27 + 1.772 =$ $.247 - .23 =$

Hint In the next problems, watch for the words **sum** and **total** for addition problems and the words **difference, how much more, how much greater**, and **how much less** for subtraction problems. In subtraction problems, be sure to put the larger number on top.

8. What is the sum of 8.9 and 14.63?

9. Find the difference between 12 and 6.488.

10. How much more is .7 than .623?

11. What is the total of 18, 2.05, and 7.3?

12. .4 is how much greater than .291?

13. Find the sum of 4.6, 2.36, and 7.

14. What is the difference between 6.1 and .61?

15. 1.47 is how much more than 1.094?

16. What is the sum of 20.2, 8.15, and .44?

17. Find the total of .73 and .483.

18. .26 is how much greater than .219?

19. What is the total of .37, .233, and .0596?

20. How much more is 18 than .9?

21. Find the sum of 3.85 and .583.

22. Find the difference between 3.85 and .583.

Answers are on page 198.

Lesson 7

MIXED ADDITION AND SUBTRACTION WORD PROBLEMS

EXERCISE

In these problems you will use either addition or subtraction. Watch for the key words or phrases, which tell you which operation to use. Then solve each problem.

1. On Monday Mrs. Ryan drove 8.5 miles to work, 5.2 miles shopping, 6.7 miles to her children's schools, and 12.3 miles back home. Find the total distance she drove that day.
   a. What is the key word or phrase?
   _____
   b. Solve the problem.

2. In 1987 the federal budget was planned to be $142.6 billion. The actual budget was $148 billion. The actual budget was how much more than the planned budget?
   a. What is the key word or phrase?
   _____
   b. Solve the problem.

Use the following information to answer questions 3 and 4.

In 1984 there were 170 million Americans of voting age. Of these, 116.1 million were registered to vote, and 101.8 million actually voted.

3. The number of people of voting age was how much greater than the number of people who registered?
   a. What is the key word or phrase?
   _____
   b. Solve the problem.

4. The number of people who registered to vote was how much greater than the number who actually voted?
   a. What is the key word or phrase?
   _____
   b. Solve the problem.

5. In a year, New York state sold $1,460 million in lottery tickets. The state made a profit of $666.8 million on the sales. What is the difference between the total sales and the profit?

   a. What is the key word or phrase?
   _____

   b. Solve the problem.

6. Oscar packed a box weighing 6.3 pounds, another weighing 4.25 pounds, and a third weighing 3.95 pounds into a large container. What was the total weight of the three boxes?

   a. What is the key word or phrase?
   _____

   b. Solve the problem.

7. In 1985 the average person in Argentina ate 178.1 pounds of beef. The average person in the United States ate 108.5 pounds of beef. How much more beef did the average Argentine eat than the average American?

   a. What is the key word or phrase?
   _____

   b. Solve the problem.

Answers are on page 198.

Lesson 8

MULTISTEP WORD PROBLEMS

EXERCISE

Each problem below requires more than one step to find the solution. The questions that follow the problems will help you decide how to find the solutions.

1. Suzanne had a fever. At 8:00 in the morning her temperature was 101.4. At noon her temperature was 1.8 higher than it was at 8:00. By midnight her temperature was 4.3 lower than her high temperature at noon. What was her temperature at midnight?

    a. The word **higher** suggests what operation, addition or subtraction? _____

    b. The word **lower** suggests what operation? _____

    c. Solve the problem.

2. The Central County school system received $2 million dollars from the state for extra programs. $.35 million was for after-school tutoring. $1.2 million was for a new adult-education center, and $.2 million was for teaching English to foreign students. How much was left over after funding these programs?

    a. What was the total amount for extra programs? _____

    b. What operation does the phrase **left over** suggest? _____

    c. Solve the problem.

3. Sam gets overtime pay when he works more than 40 hours in a week. One week he worked 8.5 hours on Monday, 6.25 hours on Tuesday, 9 hours on Wednesday, 11.5 hours on Thursday, and 9.25 hours on Friday. How many hours of overtime did Sam have that week?

    a. To find the number of hours he worked that week, do you add or subtract? _____

    b. To find the number of overtime hours, do you add to or subtract from 40? _____

    c. Solve the problem.

4. The odometer (mileage gauge) on Verna's car read 8916.4 miles on Monday morning. By Thursday night the reading was 9133.2. On Friday she drove another 85 miles. How many miles did Verna drive from Monday to Friday?

   a. To find how much she drove from Monday to Thursday, do you add or subtract? _____

   b. To find how much she drove from Monday to Friday, do you add or subtract 85? _____

   c. Solve the problem.

Answers are on page 198.

Lesson 9

TABLES

EXERCISE

Use the table below to answer the following questions.

POUNDS OF BUTTER AND MARGARINE EATEN BY AN AVERAGE AMERICAN

	1960	1970	1975	1980	1985
Butter	7.7	5.4	4.7	4.5	4.9
Margarine	9.3	10.8	11.1	11.4	10.8

SOURCE: Statistical Abstract of the U.S.

1. The average American ate how much of each of the following:
   a. butter in 1970? _____
   b. butter in 1975? _____
   c. margarine in 1960? _____
   d. margarine in 1985? _____

2. For which two years shown on the table was the amount of margarine eaten the same?

3. For which year shown on the table did the average American eat the least amount of butter?

4. For which year shown on the table did the average American eat the most butter?

5. The average American ate how much less butter in 1970 than in 1960?

6. In 1980 the average American ate how much less butter than margarine?

7. The amount of margarine the average American ate in 1985 was how much less than the 1980 amount?

Answers are on page 198.

138 Decimals

Lesson 10

BAR GRAPHS

The bar graph below tells about snowfall. Each heavy line running across the graph represents a whole number of inches. The lighter lines represent .5 inch. The short lines that do not run all the way across the graph represent .1 inch. Use the graph to answer the questions below.

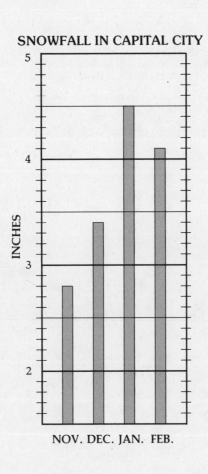

EXERCISE

1. What is the title of the graph? _____.
2. The vertical (up and down) scale is measured in _____.
3. How many months are shown on the graph? _____
4. In which month shown on the graph was there the most snow? _____
5. In which month shown on the graph was there the least snow? _____
6. The bar above January ends exactly half-way between what two whole numbers? _____
7. How much snow was there in January? _____
8. The bar above February ends how many small lines above the 4-inch line? _____

Chapter 1: Addition and Subtraction

9. How much snow was there in February? _____
10. How much snow was there in November? _____
11. How much snow was there in December? _____
12. How much more snow was there in January than in November? _____

Answers are on page 199.

Lesson 11

READING A METRIC RULER

A centimeter ruler is a tool for measuring length. It is used in most countries outside the United States. In this country the centimeter ruler is used to make precision machine parts.

The picture below shows a ruler that is 10 centimeters long. The longest lines stand for centimeters. The shortest lines stand for tenth-centimeters. These are the same as millimeters. 10 millimeters equal 1 centimeter.

To measure a length on a centimeter ruler, find how far from the left end a point is. Measurements begin at the line marked 0.

EXAMPLE 1: How far from the left end of the ruler is point *A*?

Solution: Point *A* is at the first long line on the ruler. Point *A* is 1 centimeter from the left.

EXAMPLE 2: How far from the left end of the ruler is point *B*?

Solution: Point *B* is 5 short lines to the right of 2 centimeters. Point *B* is 2.5 centimeters or 25 millimeters from the left.

EXAMPLE 3: How far from the left end of the ruler is point *C*?

Solution: Point *C* is 7 short lines to the right of 5 centimeters. Point *C* is 5.7 centimeters or 57 millimeters from the left.

To find the distance between two points on the ruler, first find how far each point is from the left end of the ruler. Then subtract these two numbers.

EXAMPLE 4: What is the distance from point *B* to point *C*?

$$\begin{array}{r} 5.7 \\ -2.5 \\ \hline 3.2 \end{array}$$ centimeters

centimeters

Solution: Point *C* is 5.7 centimeters from the left, and point *B* is 2.5 centimeters from the left. Subtract these two measurements to find the distance between them. The distance between points *B* and *C* is 3.2 centimeters.

EXERCISE

Use the 10-centimeter ruler below to answer the following questions.

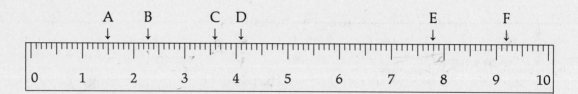

1. How far from the left end of the ruler is each point?

   a. A _____

   b. B _____

   c. C _____

   d. D _____

   e. E _____

   f. F _____

2. Tell the distance in centimeters between each of the following pairs of points.

   a. A and B _____

   b. A and C _____

   c. B and D _____

   d. C and E _____

   e. C and F _____

Answers are on page 199.

CHAPTER 2: Multiplication

Lesson 12

MULTIPLYING DECIMALS

SETTING UP PROBLEMS

You do not have to line up decimal points when you multiply. To multiply decimals, put the number with fewer nonzero digits on the bottom, and multiply as though there were no decimal points. Then count the number of decimal places in the numbers you are multiplying. Put the total number of decimal places from the two numbers in the product. Do not count a decimal point as a place.

EXAMPLE 1:   6.2 × .08 =

1	2	3	
6.2 × .08	6.2 × .08 496	6.2 × .08 .496	one decimal place two decimal places three decimal places

Step 1.   Set up the numbers for multiplication. Put .08 on the bottom.

Step 2.   Multiply 8 × 62 as though there were no decimal points.

Step 3.   Count the total number of decimal places in 6.2 and .08. There is one decimal place in 6.2 and in .08 there are two. Put three decimal places in the answer. The product is .496.

To solve the last example on a calculator, push:

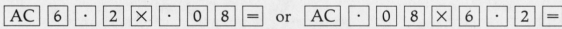

Remember that a whole number has no decimal places.

Example 2:   4.4 × 7 =

1	2	3	
4.4	4.4	4.4	one decimal place
× 7	× 7	× 7	no decimal places
	308	30.8	one decimal place

Step 1. Set up the numbers for multiplication. Put 7 on the bottom.

Step 2. Multiply 7 × 44 as though there were no decimal points.

Step 3. Count the number of decimal places in 4.4 and 7. There is one decimal place in 4.4, but in 7 there is none. Put one decimal place in the answer. The product is 30.8.

To solve the last example on a calculator, push:

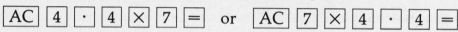

EXERCISE A

Multiply each problem.

1. .3 × .4 =   4.4 × 1.2 =   2.6 × .34 =

2. 16 × 3.07 =   3.56 × .8 =   43.2 × 8 =

3. 6.8 × 22 =   .59 × 7.3 =   46 × .7 =

Answers are on page 199.

ADDING ZEROS

Sometimes you have to add zeros to the product.

Example 3:   .5 × .09 =

1	2	3	
.09	.09	.09	two decimal places
× .5	× .5	× .5	one decimal place
	45	.045	three decimal places

Step 1. Set up the numbers for multiplication. Put .5 on the bottom.

Step 2. Multiply 9 × 5 as though there were no decimal points.

Step 3. Count the decimal places in .09 and .5. There are two decimal places in .09 and in .5 there is one. You need three decimal places in the product. Put a zero to the left of 45 to make three decimal places. The product is .045.

 To solve the last example on a calculator, push:

| AC | . | 0 | 9 | × | . | 5 | = | or | AC | . | 5 | × | . | 0 | 9 | = |

EXERCISE B

Multiply each problem.

1. 1.2 × .06 =   .46 × .08 =   .3 × .015 =

2. .006 × 14 =   .003 × .72 =   2.4 × .03 =

Answers are on page 199.

CANCELING ZEROS

Sometimes you can cancel zeros in the product.

EXAMPLE 4:   4.5 × .6 =

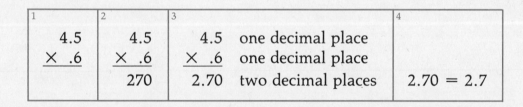

1	2	3	4
4.5 × .6	4.5 × .6 270	4.5 one decimal place × .6 one decimal place 2.70 two decimal places	2.70 = 2.7

Step 1. Set up the numbers for multiplication. Put .6 on the bottom.

Step 2. Multiply 6 × 45 as though there were no decimal points.

Step 3. Count the decimal places in 4.5 and .6. There is one decimal place in 4.5 and in .6 there is also one. The product needs two decimal places, 2.70.

Step 4. The last zero in 2.70 has no use as a placeholder. Simplify the answer to 2.7.

 To solve the last example on a calculator, push:

| AC | 4 | . | 5 | × | . | 6 | = | or | AC | . | 6 | × | 4 | . | 5 | = |

EXERCISE C

Multiply each problem.

1. .5 × .4 =   1.35 × 20 =   .95 × 8 =

2. 2.6 × .45 =   1.5 × 3.6 =   42 × 2.5 =

3. 3.4 × 1.3 =   9 × 6.2 =   .3 × 12.9 =

4. 8 × .02 =   .55 × 6 =   1.26 × .3 =

5. 4.5 × 15.4 =   2.1 × .1 =   34 × .7 =

6. 29 × 1.6 =   4.86 × 4 =   .6 × .055 =

7. 3.78 × .4 =   .76 × .15 =   5 × 3.46 =

8. .09 × .17 =   7.3 × .01 =   48 × .13 =

9. 18 × 1.5 =   .005 × 1.6 =   129 × .001 =

10. 10.2 × .25 =   .64 × 135 =   3.8 × 65 =

Answers are on page 199.

Lesson 13

MULTIPLYING BY 10, 100, AND 1000

Multiplying a decimal or a mixed decimal by 10, 100, or 1000 is as easy as moving a decimal point.

To multiply a decimal by 10, move the decimal point one place to the right.

Example: $4.23 \times 10 = 42.3$.

To multiply a decimal by 100, move the decimal point two places to the right.

Example: $38.2 \times 100 = 3820$.

Notice that you must add a zero to 38.2 to move the point two places.

To multiply a decimal by 1000, move the decimal point three places to the right.

Example: $.45 \times 1000 = 450$.

Notice, again, that you must add a zero to .45 to get enough decimal places.

EXERCISE

Multiply each problem.

1. $1.4 \times 100 =$ $6.8 \times 1000 =$ $5.3 \times 10 =$

2. $.3 \times 1000 =$ $.47 \times 10 =$ $.7 \times 100 =$

3. $.28 \times 10 =$ $.365 \times 100 =$ $.18 \times 1000 =$

4. $16.4 \times 100 =$ $8.7 \times 1000 =$ $4.25 \times 10 =$

5. $1.28 \times 1000 =$ $7.333 \times 10 =$ $27.46 \times 100 =$

6. $5.2 \times 10 =$ $1.3 \times 100 =$ $30.5 \times 1000 =$

7. Find the weight of 100 identical packages if each package weighs 1.2 pounds. _____

8. The distance from Robert's house to the factory where he works is .8 mile. He makes this trip 10 times a week. Find the total distance he drives to and from work each week. _____

Answers are on page 199.

Lesson 14

ROUNDING OFF DECIMALS

$2.48 is closer to $2.50 than to $2.40. **Rounded off** to the nearest ten cents, $2.48 is $2.50.

To round off a decimal, follow these steps:

- Underline the digit in the place you want to round off to.
- If the digit to the right of the underlined digit is more than 4, add 1 to the underlined digit and drop the numbers to the right.

EXAMPLE 1:   Round off .294 to the nearest tenth.

1	2
.2̲94	.3

Step 1.   Underline the digit in the tenths place, 2.

Step 2.   Since 9 is to the right of 2 and 9 is more than 4, add 1 to 2. Rounded off to the nearest tenth, .294 is .3. Notice that the answer has only one decimal place since you were asked to round off to tenths.

- If the digit to the right of the underlined digit is less than 5, leave the underlined digit as it is and drop the numbers to the right.

EXAMPLE 2:   Round off .6738 to the nearest hundredth.

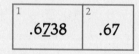

Step 1.   Underline the digit in the hundredths place, 7.

Step 2.   Since 3 is to the right of 7 and 3 is less than 5, leave 7 as it is. .6738 rounded off to the nearest hundredth is .67. Notice that the answer has only two places since you were asked to round off to hundredths.

EXERCISE

1. Round off each number to the nearest tenth or ten cents.
   1.36   27.345   .462 $5.119 $8.4369

2. Round off each number to the nearest hundredth or penny.
   1.385   .2962   1.386 $2.8762 $.3072

3. Round off each number to the nearest thousandth.
   .2253   8.4235   .3333   10.5974   2.8855

4. Round off each number to the nearest unit or dollar.
   3.449   16.5   8.258 $5.72 $19.56

5. One September, 6.56 inches of rain fell in Mobile, Alabama. Round off the number to the nearest tenth. _____

6. Round off the amount of rain in the last problem to the nearest unit (inch). _____

Answers are on page 199.

Lesson 15

MULTIPLICATION WORD PROBLEMS

EXERCISE

In most multiplication problems you have information about one thing, and you must find information about several things. The questions after the problems will help you find the information about one thing.

1. The classes at the Capital City adult school meet for 16 weeks. The classes meet 6.5 hours each week. If a student attends every class, how many hours does he or she spend altogether?

   a. What is the number of class **hours per week**? _____

   b. Solve the problem.

2. Phil is mailing 15 announcements of his daughter's birth. Each announcement weighs 1.3 ounces. Find the total weight of the announcements.

   a. What is the **weight per announcement**?

   _____

   b. Solve the problem.

3. Sandra makes $5.35 an hour. How much does she make for working 37.5 hours? Round off the answer to the nearest penny.

   a. What is Sandra's **wage per hour**?

   _____

   b. Solve the problem.

4. Jim drove 1.5 hours at a speed of 36 miles an hour. How far did he drive?

   a. What was Jim's speed in **miles per hour**?

   _____

   b. Solve the problem.

5. One gallon of gasoline costs $1.089. To the nearest penny, how much did Lois pay for 13 gallons of gasoline?

   a. What was the price of the gasoline in **dollars per gallon**? _____

   b. Solve the problem.

150 *Decimals*

6. What is the cost of 3.25 pounds of tomatoes if the tomatoes cost $.78 a pound? Round off to the nearest penny.
   a. What was the price of the tomatoes in **dollars per pound**? _____
   b. Solve the problem.

7. An inch contains 2.54 centimeters. 12 inches contain how many centimeters?
   a. There are how many **centimeters per inch**? _____
   b. Solve the problem.

Answers are on page 199–200.

Lesson 16

MIXED OPERATION WORD PROBLEMS

EXERCISE

Each of the following problems requires more than one operation. The questions that follow each problem will help you decide how to find the solutions.

1. The Dumonts drove for 4.5 hours at an average speed of 60 mph and then for 1.5 hours at an average speed of 32 mph. Find the total distance they drove.
   a. How far did they drive at 60 mph?
   _____
   b. How far did they drive at 32 mph?
   _____
   c. To find the total distance, do you add or subtract?
   _____
   d. Solve the problem.

2. Mrs. Brown bought 3.5 pounds of chicken at $1.49 a pound. How much change should she get from $10?
   a. What was the cost per pound for the chicken?
   _____
   b. What was the total cost of the chicken?
   _____
   c. To find change means to add or subtract?
   _____
   d. Solve the problem.

3. One week Jack worked for 40 hours at $5.50 an hour and for 6.5 hours of overtime at $8.25 an hour. Altogether how much did he make that week?
   a. How much did he make for the first 40 hours?
   _____
   b. How much did he make for the overtime hours?
   _____

152 Decimals

c. The word **altogether** suggests what operation?
_____

d. Solve the problem.

4. Pete is building a bookshelf. He needs 12.5 yards of lumber, which costs $6.80 a yard. He will charge $75 for his labor. What is the total cost of the bookshelf?

   a. What is the cost of the lumber?
   _____

   b. Do you need to add or subtract the labor costs?
   _____

   c. Solve the problem.

5. Calvin has 24.5 acres of his farm planted in soybeans. He gets 30 bushels of soybeans for each acre. He gets paid $4.60 a bushel for soybeans. Find the total value of the soybeans.

   a. How many bushels of soybeans does Calvin produce?
   _____

   b. What is the price per bushel?
   _____

   c. Solve the problem.

Answers are on page 200.

Lesson 17

PERIMETER AND AREA OF RECTANGLES

On page 78 you learned that perimeter is the distance around a flat figure. The formula to find the perimeter of a rectangle is $P = 2L + 2W$. Remember that perimeter is measured in units, such as yards or meters.

On page 80 you learned that area is the amount of surface on a flat figure. The formula to find the area of a rectangle is $A = LW$. Remember that area is measured in **square** units, such as square feet or square meters.

EXERCISE

Find both the perimeter and the area of each figure below.

1.   7 feet / 3.5 feet

2.   6.6 meters / 4.2 meters

3.   3.2 inches / 5.4 inches

4.   16 yards / 7.5 yards

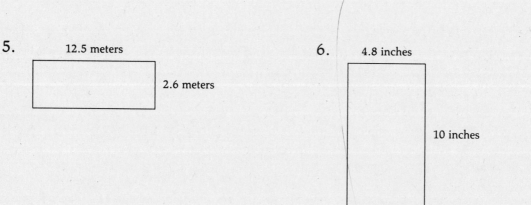

5.   12.5 meters / 2.6 meters

6.   4.8 inches / 10 inches

Answers are on page 200.

154   Decimals

Lesson 18

Circles

A **circle** is a curved, flat figure. Every point on the circle is the same distance from the center. Pizzas, can lids, and paper plates are usually shaped like circles. On this page you will learn several of the special terms that describe a circle.

Circumference

The distance around a circle is called the **circumference**. The circumference of a circle is like the perimeter of a rectangle.

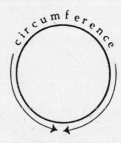

Radius

The distance from the center of a circle to the outside is the **radius**. A radius is like a spoke on a bicycle wheel.

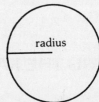

Diameter

The widest distance across a circle is the **diameter**. The diameter of a circle is twice as long as the radius. The diameter goes through the center of a circle.

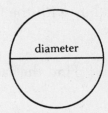

Pi

A special number known as π (the Greek letter **pi**) appears in the formulas you will see on the following pages. π is a number that shows the relationship between the circumference and the diameter of any circle. π does not have an exact value. However, the decimal 3.14 is close.

FINDING THE DIAMETER

The diameter of a circle is two times the radius. If you know the radius of a circle, you can find the diameter with the formula $d = 2r$, where d is the diameter and r is the radius.

EXAMPLE: Find the diameter of the circle pictured at the right.

$d = 2r$
$d = 2 \times 1.5 = 3$ feet

Solution: Replace r with 1.5 in the formula $d = 2r$. Then multiply 2×1.5. The radius is 3.0 or 3 feet.

Chapter 2: Multiplication 155

EXERCISE A

Find the diameter (d) of each circle pictured below.

1.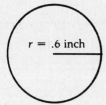

2.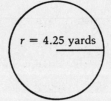

3.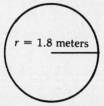

4.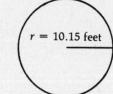

Answers are on page 200.

FINDING THE RADIUS

The radius is half the diameter. If you know the diameter of a circle, you can find the radius with the formula $r = .5d$, where r is the radius and d is the diameter.

EXAMPLE: Find the radius of the circle pictured at the right.

$r = .5d$
$r = .5 \times 8 = 4$ yards

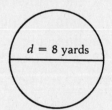

Solution: Replace d with 8 in the formula $r = .5d$. Then multiply 0.5×8. The diameter is $4.0 = 4$ yards.

EXERCISE B

Find the radius (r) of each circle pictured below.

1.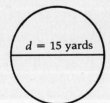

2.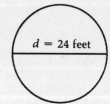

3.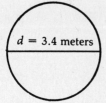

4.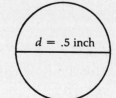

Answers are on page 201.

Lesson 19

CIRCUMFERENCE OF A CIRCLE

The **circumference** is the distance around a circle. Circumference is measured in units, such as inches or meters.

To find the circumference of a circle, use the formula $C = \pi d$.

C is the circumference.

π is 3.14.

d is the diameter.

EXAMPLE: Find the circumference of the circle pictured at the right.

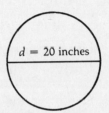

$C = \pi d$
$C = 3.14 \times 20 = 62.8$ inches

Solution: Replace π with 3.14 and d with 20 in the formula $C = \pi d$. Then multiply 3.14×20. The circumference of the circle is 62.80 or 62.8 inches.

To solve the last example on a calculator, push:

| AC | 3 | . | 1 | 4 | × | 2 | 0 | = |

EXERCISE

Find the circumference of each of the following. Remember to write the correct unit, such as feet or meters, with each answer.

1.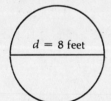

2.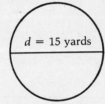

3.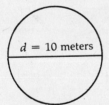

4.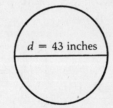

5. What is the circumference of a circular pool with a diameter of 50 feet?

6. Find the circumference of a round table top with a diameter of 65 inches. Round off to the nearest inch.

Answers are on page 201.

Lesson 20

AREA OF A CIRCLE

The area of a circle is the amount of surface inside the circle. Area is measured in square units, such as square feet or square miles.

To find the area of a circle, use the formula $A = \pi r^2$.

A is the area.

π is 3.14.

r is the radius.

The small 2 is called an **exponent**. It means to write the radius two times and to multiply. r^2 means r times r.

EXAMPLE: Find r^2, when r is 6.

$$r^2 = 6^2 = 6 \times 6 = 36.$$

Solution: Replace r with 6. Then write 6 two times and multiply. $6^2 = 36$.

EXERCISE A

Solve each of the following.

1. $7^2 =$
2. $12^2 =$
3. $9^2 =$
4. $1.3^2 =$

Answers are on page 201.

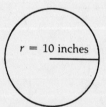

EXAMPLE: Find the area of the circle pictured at the right.

$A = \pi r^2$
$A = 3.14 \times 10^2$

$A = 3.14 \times 10 \times 10 = 314$ square inches

Step 1. Replace π with 3.14 and r with 10 in the formula $A = \pi r^2$.

Step 2. Write 10^2 as 10×10 and multiply across. The area is 314 square inches.

To solve the last example on a calculator, push:

| AC | 3 | . | 1 | 4 | × | 1 | 0 | × | 1 | 0 | = |

EXERCISE B

Find the area of each of the following circles:

1.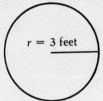

2.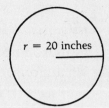

3.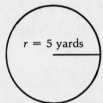

4.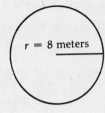

5. Find the area of a circular dance floor that has a radius of 30 feet.

6. Find the area of the bottom of a circular pool that has a radius of 4 yards.

Answers are on page 201.

Chapter 2: Multiplication   159

CHAPTER 3: Division

Lesson 21

DIVIDING DECIMALS BY WHOLE NUMBERS

DIVIDING

To divide a decimal by a whole number, divide as you would with whole numbers. Then bring the decimal point up into the quotient above its position in the dividend.

EXAMPLE 1: $1.2 \div 3 =$

```
   1              2
         4              .4
      ─────          ─────
      3)1.2          3)1.2
       1 2            1 2
       ───            ───
         0              0
```

Step 1. Divide 12 by 3.

Step 2. Bring the decimal point up into the quotient above its position in the dividend. The quotient is .4.

On a calculator, remember to punch in the dividend first. To solve the last example on a calculator, push:

[AC] [1] [.] [2] [÷] [3] [=]

ZEROS

Be careful with zeros.

EXAMPLE 2: $.45 \div 5 =$

```
   1              2
         9             .09
      ─────          ─────
      5).45          5).45
        45             45
        ──             ──
         0              0
```

160 Decimals

Step 1. Divide 45 by 5.

Step 2. Bring the decimal point up into the quotient above its position in the dividend. You must write a zero in the tenths place. This zero shows that the digit 9 is in the hundredths place.

To solve the last example on a calculator, push:

$\boxed{AC}$ $\boxed{\cdot}$ $\boxed{4}$ $\boxed{5}$ $\boxed{\div}$ $\boxed{5}$ $\boxed{=}$

EXERCISE

Divide each problem.

1. $10.4 \div 4 =$ $1.62 \div 9 =$ $40.6 \div 7 =$

2. $3.95 \div 5 =$ $.216 \div 6 =$ $50.4 \div 8 =$

3. $1.12 \div 16 =$ $8.16 \div 34 =$ $2.196 \div 61 =$

4. $24.08 \div 56 =$ $.675 \div 75 =$ $195.5 \div 23 =$

5. $80.73 \div 39 =$ $2846.9 \div 83 =$ $4.922 \div 46 =$

6. $598.6 \div 73 =$ $1.273 \div 19 =$ $22.04 \div 38 =$

Answers are on page 201.

Lesson 22

DIVIDING DECIMALS BY DECIMALS

MOVING DECIMAL POINTS

To divide by a decimal, first change the problem. Make the divisor a whole number by moving the decimal point as far to the right as it will go. Then move the decimal point in the dividend the **same** number of places that you moved the point in the divisor. These moves make a new problem that is equal to the original problem.

EXAMPLE 1:   2.24 ÷ .7 =

1	2	3
.7.)2.2.4	3 2 .7.)2.2.4 2 1 1 4 1 4 0	3.2 .7.)2.2.4 2 1 1 4 1 4 0

Step 1. Move the decimal point in the divisor, .7, one place to the right to make it a whole number. Then move the decimal point in the dividend one place to the right.

Step 2. Divide 224 by 7.

Step 3. Bring the decimal point up into the quotient above its new position in the dividend. The quotient is 3.2.

To solve the last example on a calculator, push:

[AC] [2] [.] [2] [4] [÷] [.] [7] [=]

EXAMPLE 2:   .027 ÷ .03 =

1	2	3
.03.).02.7	9 .03.).02.7 2 7 0	.9 .03.).02.7 2 7 0

162   *Decimals*

Step 1. Move the decimal point in the divisor, .03, two places to the right to make it a whole number. Then move the decimal point in the dividend two places to the right.

Step 2. Divide 3 into 27.

Step 3. Bring the decimal point up into the answer above its new place in the dividend. The quotient is .9.

To solve the last problem on a calculator, push:

| AC | . | 0 | 2 | 7 | ÷ | . | 0 | 3 | = |

ZEROS

You may have to add one or more zeros to the dividend.

EXAMPLE 3: 12.8 ÷ .16 =

Step 1. Move the decimal point in the divisor, .16, two places to the right to make it a whole number. To move the decimal point in the dividend two places to the right, first put a zero at the right of 12.8. Then move the point two places.

Step 2. Divide 1280 by 16.

Step 3. Bring the decimal point up into the answer above its new place in the dividend. The quotient is 80.

To solve the last example on a calculator, push:

| AC | 1 | 2 | . | 8 | ÷ | . | 1 | 6 | = |

Chapter 3: Division 163

EXERCISE

Divide each problem.

1. $19.2 \div .8 =$ $5.2 \div .4 =$ $32.4 \div .9 =$

2. $.85 \div .5 =$ $2.94 \div .7 =$ $11.4 \div .06 =$

3. $.704 \div 3.2 =$ $4.08 \div 1.2 =$ $.024 \div 1.6 =$

4. $.5131 \div .07 =$ $2.66 \div .019 =$ $.1184 \div .08 =$

5. $18.48 \div .6 =$ $1.095 \div .5 =$ $38.25 \div .009 =$

6. $31.5 \div .15 =$ $62.1 \div .27 =$ $.00208 \div .13 =$

7. $19.35 \div 4.5 =$ $13.44 \div .056 =$ $.3496 \div .92 =$

8. $38.69 \div .53 =$ $1.458 \div .027 =$ $16.47 \div 6.1 =$

9. $.708 \div .12 =$ $.5208 \div 8.4 =$ $38.18 \div .046 =$

10. $213.18 \div 6.6 =$ $11.256 \div .14 =$ $13.376 \div 3.2 =$

11. $2.5871 \div .041 =$ $197.1 \div 7.3 =$ $.09108 \div .018 =$

12. $100.1 \div 1.43 =$ $.01944 \div .216 =$ $.06832 \div 4.27 =$

13. $6.72 \div 1.68 =$ $1.2029 \div .523 =$ $1227.6 \div 6.82 =$

Answers are on page 201.

Lesson 23

DIVIDING WHOLE NUMBERS BY DECIMALS

MOVING DECIMAL POINTS

The decimal point is at the right in any whole number. When you divide a whole number by a decimal, put a decimal point at the right of the whole number. Move the decimal point in the divisor to make it a whole number. Then move the decimal point in the dividend the same number of places.

EXAMPLE: 72 ÷ .08 =

1	2	3
.08.)72.00.	9 00 .08.)72.00. 72 0 00	9 00. .08.)72.00. 72 0 00

Step 1. Move the decimal point in the divisor, .08, two places to the right to make it a whole number. Put a decimal point at the right of 72. To move the decimal point in the dividend, 72, two places to the right, you must add two zeros.

Step 2. Divide 7200 by 8.

Step 3. Bring the decimal point up into the answer above its new position in the dividend. The quotient is 900.

 To solve the last problem on a calculator, push:
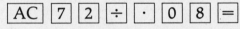

EXERCISE

Divide each problem.

1. 36 ÷ .9 =   42 ÷ .6 =   28 ÷ .04 =

2. 750 ÷ 1.5 =   78 ÷ 1.3 =   96 ÷ 1.2 =

3. 720 ÷ .24 =   860 ÷ 4.3 =   63 ÷ .07 =

4. 910 ÷ 1.3 =   1680 ÷ .21 =   72 ÷ .018 =

5. 112 ÷ .7 =   243 ÷ .09 =   344 ÷ .8 =

6. 128 ÷ .04 =   144 ÷ .6 =   405 ÷ .5 =

Answers are on page 201.

Lesson 24

MIXED DIVISION PROBLEMS

EXERCISE

These problems give you a chance to practice the division of decimals skills you have learned so far. Divide each problem.

1. 7.8 ÷ .13 =   2.16 ÷ .24 =   63 ÷ .9 =

2. .624 ÷ .12 =   .036 ÷ 1.8 =   .108 ÷ .36 =

3. 208 ÷ 2.6 =   1.216 ÷ 38 =   3.01 ÷ 43 =

4. 15.36 ÷ .64 =   105 ÷ .15 =   3.78 ÷ .21 =

5. 30.68 ÷ 5.9 =   22.05 ÷ .63 =   3956 ÷ 9.2 =

6. 1250 ÷ 2.5 =   .1008 ÷ .016 =   360 ÷ .45 =

7. 1.625 ÷ 1.25 =   2.205 ÷ 63 =   54 ÷ .6 =

8. .91 ÷ 35 =   90 ÷ .015 =   3.44 ÷ 43 =

Answers are on page 201.

Lesson 25

Uneven Decimal Division

ROUNDING ANSWERS OFF

Some decimal division problems never come out even, no matter how many zeros you add. You can round off the answer by dividing to one place beyond the place you want to round off to. For example, if you want an answer to the nearest hundredth, divide to the thousandths place. Look at page 25 for rounding off.

Example: Round off the answer to .8 ÷ 3.5 to the nearest hundredth.

1	2	3

3.5.)̅.8̲.̲0̲0̲0̲	.228
	3.5.)̅.8̲.̲0̲0̲0̲
	   7 0
	   1 00
	     70
	    300
	    280
		to the nearest hundredth = .23

Step 1. Move the decimal point one place to the right in both the divisor and the dividend. Since you want to round off to hundredths, write three zeros to the right of the new decimal point in the dividend. This gives you thousandths, which is one more place than hundredths.

Step 2. Divide 8000 by 35.

Step 3. Round off .228 to the nearest hundredth. The answer is .23.

To solve the last problem on a calculator, push:

[AC] [.] [8] [÷] [3] [.] [5] [=]

Notice that the calculator fills all the spaces on the answer screen. The answer on the calculator is 0.2285714. To round off to hundredths, you need only 0.228.

EXERCISE

1. Divide each problem. Round off each answer to the nearest tenth.

   a. $4.4 \div .6 =$

   b. $6.8 \div 1.2 =$

   c. $4.2 \div 9 =$

2. Divide each problem. Round off each answer to the nearest hundredth.

   a. $4 \div .15 =$

   b. $2.2 \div 17 =$

   c. $5.2 \div .09 =$

3. Divide each problem. Round off each answer to the nearest thousandth.

   a. $.01 \div .007 =$

   b. $7 \div 1.2 =$

   c. $.35 \div 18 =$

Answers are on page 201.

Lesson 26

DIVIDING BY 10, 100, AND 1000

Dividing a decimal by 10, 100, or 1000 is as easy as moving a decimal point. To divide a decimal by 10, move the decimal point one place to the left.

Example: 3.8 ÷ 10 = .38.

To divide a decimal by 100, move the decimal point two places to the left.

Example: .2 ÷ 100 = .002.

You must add zeros to get two more decimal places.
To divide a decimal by 1000, move the decimal point three places to the left.

Example: 8.2 ÷ 1000 = .0082.

EXERCISE

Divide each problem.

1. 2.3 ÷ 100 =   1.5 ÷ 1000 =   .3 ÷ 10 =

2. .16 ÷ 1000 =   .327 ÷ 10 =   .55 ÷ 100 =

3. 18.9 ÷ 10 =   .2 ÷ 100 =   1.9 ÷ 1000 =

4. 25 ÷ 100 =   12.6 ÷ 1000 =   4.832 ÷ 10 =

5. 13 ÷ 1000 =   54 ÷ 10 =   9 ÷ 100 =

6. .04 ÷ 10 =   2.1 ÷ 100 =   .7 ÷ 1000 =

7. A pipe 2.3 meters long was cut into 10 equal pieces. How long was each piece? _____

8. 125 pounds of berries were divided up evenly into 100 boxes. What was the weight of the berries in each box? _____

Answers are on page 201.

Lesson 27

DIVISION WORD PROBLEMS

EXERCISE

In every division problem, remember to put the amount being divided, the dividend, inside the $\sqrt{}$ sign. On a calculator, be sure to enter the dividend first. The questions after each problem should help you find the dividend.

1. Mr. Parrish split 29.2 kilograms of apples evenly among four people. How many kilograms did each person get?
   a. What is being divided up, the apples or the people? _____
   b. Solve the problem.

2. Eric earns $306 for a 40-hour week. How much does Eric make in one hour?
   a. How is the answer measured, in dollars or hours? _____
   b. What is being divided up, the dollars or the hours? _____
   c. Solve the problem.

3. Sal bought 12.5 gallons of gasoline for a total of $13. How much did he pay for one gallon?
   a. How is the answer measured, in dollars or gallons? _____
   b. What is being divided up, the dollars or the gallons? _____
   c. Solve the problem.

4. Mr. and Mrs. Robinson, their daughter Cheryl, and Cheryl's daughter Adrienne plan to drive 250 miles. If Cheryl and her parents share the driving equally, how many miles will each of them drive? Round the answer off to the nearest mile.
   a. What is being divided, the miles or the people? _____
   b. How many people are sharing the driving? _____
   c. Solve the problem.

5. Veronica agrees to pay $49.50 a month for new furniture. In how many months can she pay off the total of $990 that she owes?

   a. What is being divided up, the total bill of $990 or the monthly payment of $49.50? _____

   b. Solve the problem.

6. A 6.5-ounce can of tuna costs $1.19. Find, to the nearest penny, the price of one ounce of tuna.

   a. What is being divided up, the price or the weight of the can?
   _____

   b. Solve the problem.

Answers are on page 202.

Lesson 28

MIXED OPERATION WORD PROBLEMS

EXERCISE

Each of the following problems requires more than one operation. Following each problem are questions that will help you decide how to solve the problems.

1. Phil picked 43.5 pounds of peaches. Sue picked 53 pounds, and Ann picked 26.5 pounds. They agreed to share the peaches equally.

    a. How many pounds of peaches did they pick altogether? _____

    b. How many people are sharing the peaches? _____

    c. How many pounds did each of them get? _____

2. Carmen works part time at a grocery store. On Monday she worked 3.5 hours. On Thursday she worked 4.5 hours, and on Saturday she worked 5.5 hours. Altogether she made $56.70. How much does she get for one hour of work?

    a. Altogether how many hours did she work? _____

    b. How is the final answer measured, in dollars or hours? _____

    c. What is being divided up, dollars or hours? _____

    d. Solve the problem.

3. Together Mary and David make $19,344 a year. They decide that they can afford no more than .25 times their income for rent. What is the most **monthly** rent they can pay?

    a. The phrase **.25 times** suggests what operation? _____

    b. If you know a yearly amount, how do you find the amount for one month?
    _____

Chapter 3: Division 173

There are two ways to solve this problem. The next questions will help you think through both solutions.

c. What is the most **yearly** rent they can afford? _____

d. Based on the last answer, what is the monthly rent they can pay?
_____

e. What is their **monthly** income?
_____

f. Based on the last answer, what is the monthly rent they can pay?
_____

4. Paul drove 18 miles to work, 16.7 miles after work to watch his son's baseball game, and 14.5 miles back home. He used three gallons of gasoline. How far did he drive that day on one gallon of gasoline?

   a. What total distance did he drive that day? _____

   b. How is the final answer measured, in miles or gallons? _____

   c. Solve the problem.

Answers are on page 202–203.

174 *Decimals*

Lesson 29

AVERAGE

EXERCISE

You learned on page 106 that an average is a total divided by the number of items in the total. Solve each of the following decimal problems.

1. In June there were 4.7 inches of rain. In July there were 3.3 inches of rain, and in August there were 2.5 inches. What was the average rainfall for this period?

2. The Hartmans drove 86.2 miles on Friday, 306.9 miles on Saturday, and 230.3 miles on Sunday. Find the average number of miles they drove each day.

3. Frank mailed a package that weighed 7.25 pounds, another that weighed 3.4 pounds, and a third that weighed 8.1 pounds. What was the average weight per package?

4. A gallon of gasoline at the corner station costs $1.089. One block away another station charges $1.149. What is the average of these prices? Round your answer off to the nearest penny.

5. Following are the 11:00 A.M. temperature readings for five days during a week in August: 74°, 78°, 83°, 86°, and 82°. Find the average of these temperature readings. Round your answer off to the nearest whole number.

6. Following are the batting averages of the four best hitters of the Springfield Hawks: .435, .325, .316, and .290. What is the average of these batting averages? Round your answer off to the nearest thousandth.

Answers are on page 203.

Lesson 30

Gas Mileage

Some drivers keep track of the amount of gasoline they buy. Gas mileage is the rate a car, or other vehicle, burns up fuel. Gas mileage is measured in miles per gallon, or mpg. Miles per gallon tells you the distance you travel on one gallon of gas. To find gas mileage, divide the number of miles traveled by the number of gallons of gasoline used.

Example: Find the gas mileage for a car that travels 70 miles on 3.5 gallons of gasoline.

```
            2 0  mpg
      ┌────────
3.5.)│ 7 0.0.
            7 0
            ───
            0 0
```

Solution: Divide 70 miles by 3.5 gallons. The gas mileage is 20 mpg.

To solve the last example on a calculator, push:

[AC] [7] [0] [÷] [3] [·] [5] [=]

Hint If you multiply **miles per gallon** by the number of gallons a driver uses, you will find the distance traveled. For the example, 20 × 3.5 = 70.0 = 70 miles.

Exercise

Find the gas mileage for each of the following:

1. A car that drives 150 miles on 8.5 gallons of gasoline. Round your answer to the nearest whole number.

2. A car that drives 100 miles on 9 gallons of gasoline. Round your answer to the nearest whole number.

3. A car that drives 170 miles on 12.5 gallons of gasoline. Round your answer to the nearest whole number.

4. A bus that travels 260 miles on 45 gallons of gasoline. Round your answer to the nearest tenth.

5. A bus that travels 145 miles on 30.5 gallons of gasoline. Round your answer to the nearest tenth.

6. A truck that travels 300 miles on 35 gallons of gasoline. Round your answer to the nearest tenth.

7. A truck that travels 460 miles on 50 gallons of gasoline. Round your answer to the nearest whole number.

Answers are on page 203.

Lesson 31

BATTING AVERAGES

A batting average tells how well a baseball player is hitting. Batting averages are measured in thousandths. When a baseball fan says a player is batting "three hundred," he means the player hits 300 times out of 1000 times at bat. This is the same as 30 out of every 100 times or 3 out of every 10 times.

To find a batting average, divide the number of hits a player gets by the number of times the player is at bat. The answer is always expressed in thousandths. Divide to the ten-thousandths place, then round off to thousandths.

EXAMPLE: Gordon was at bat 85 times, and he got 30 hits. What was his batting average?

1	2
$$\begin{array}{r}.3529\\85\overline{)30.0000}\\\underline{25\ 5}\\4\ 50\\\underline{4\ 25}\\250\\\underline{170}\\800\\\underline{765}\end{array}$$	to the nearest thousandth = .353

Step 1. Divide the number of hits, 30, by the number of times at bat, 85. Put four zeros for ten-thousandths in the problem.

Step 2. Round off to the nearest thousandth. Gordon's batting average was .353.

To solve the last problem on a calculator, push:

[AC] [3] [0] [÷] [8] [5] [=]

178 *Decimals*

EXERCISE

Find the batting average for each player below.

Player	Hits	At Bat	Average	
			(ten-thousandths)	(thousandths)
Randolph	26	80		
Gomez	15	58		
Smith	10	23		
Rigby	20	69		
Johnson	50	158		
Howard	15	73		
Acevedo	24	144		

Answers are on page 204.

DECIMALS REVIEW

These problems will help you decide what you need to review about decimals.

1. .8 + .5 =
2. .43 + .276 =

3. 3.9 + 2.76 + .48 =
4. 16 + 2.935 + 1.8 =

5. .7 − .1 =
6. .2 − .137 =

7. 8 − 3.06 =
8. 14.3 − 7.667 =

9. .4 × .7 =
10. 3.06 × 9 =

11. 1.365 × 4.2 =
12. .019 × .8 =

13. 2.43 ÷ 9 =
14. .224 ÷ .16 =

15. .288 ÷ 4.8 =
16. 9 ÷ .015 =

17. Write six tenths as a decimal. _____

18. Write five and four hundredths as a mixed decimal. _____

19. Write fourteen thousandths as a decimal. _____

20. Which is larger, .3 or .19? _____

21. Which is larger .08 or .096? _____

22. Round off 4.236 to the nearest tenth. _____

23. Round off .148 to the nearest hundredth. _____

24. Round off 6.547 to the nearest unit. _____

25. When Rachel was sick, her temperature went up 3.5° from her normal temperature of 98.6°. What was her high temperature?

26. To build a new city hall, the town of Palmdale hopes to raise $3 million. So far they have $1.77 million. How much more money do they need?

27. Larry usually works 37.5 hours a week. One week he worked 3.25 hours overtime on Wednesday, 2.5 hours overtime on Thursday, and 4.5 hours overtime on Friday. How many hours did he work altogether that week?

28. The odometer (mileage gauge) on Don's car read 7428.2 on Friday morning. By Sunday morning it read 7634.1. By Sunday night Don had driven another 115 miles. How far did Don drive altogether from Friday morning until Sunday night?

Use the table below to answer questions 29 to 32 on the next page.

NUMBER OF LONG-PLAYING RECORDS AND CASSETTES SOLD IN THE U.S. (IN MILLIONS)

	1981	1982	1983	1984	1985	1986
LP records	295.2	243.9	209.6	204.6	167.0	125.2
Cassettes	137.0	182.3	236.8	332.0	339.1	344.5

SOURCE: Statistical Abstract of the U.S.

29. What was the first year shown on the table when more cassettes were sold than long-playing records?

30. How many more cassettes were sold in 1986 than in 1981?

31. The sales of long-playing records dropped by how many from 1985 to 1986?

32. The number of long-playing records sold in 1981 was how much more than the **combined** number of records sold in both 1985 and 1986?

Use the centimeter ruler below to answer questions 33 to 36.

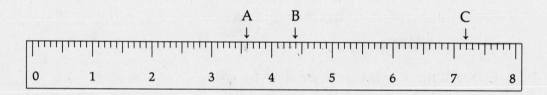

33. How far from the left end of the ruler is point *A*? _____

34. How far from the left end of the ruler is point *C*? _____

35. What is the distance from point *A* to point *B*?

36. What is the distance from point *A* to point *C*?

---

37. Mitchell makes $6.35 an hour. To the nearest penny, how much does he make for 7.5 hours of work?

38. Ann drove for 2.5 hours at an average speed of 52 mph. How far did she drive?

39. What is the cost of 4.25 pounds of potatoes at $.18 a pound? Round your answer to the nearest penny.

40. An inch is 2.54 centimeters. 10 inches is equal to how many centimeters?

41. The formula for the area of a rectangle is $A = LW$, where A is the area, L is the length, and W is the width. Find the area of a rectangle that is 4.6 meters long and 3.4 meters wide.

42. The formula for the circumference of a circle is $C = \pi d$, where C is the circumference, $\pi = 3.14$, and d is the diameter. What is the circumference of a circle with a diameter of 12 feet?

43. The formula for the area of a circle is $A = \pi r^2$, where A is the area, $\pi = 3.14$, and r is the radius. What is the area of a circle with a radius of 30 feet?

44. Jerry makes $234 for 40 hours of work. How much is his hourly wage?

45. What is the average weight per box of three boxes that weigh 5.2 pounds, 4.95 pounds, and 2.6 pounds?

46. Mike drove 120 miles on 11 gallons of gasoline. To the nearest tenth of a mile, how many miles did he drive on one gallon of gasoline?

47. A batting average is the number of hits a baseball player makes divided by the number of times he is at bat. Bill was at bat 24 times, and he made 7 hits. What was his batting average to the nearest thousandth?

48. Mark and Heather make $22,380 a year. Their food budget is .3 times their income. Find their **monthly** food budget.

49. Miriam made $33 for 7.5 hours of work. How much did she make in one hour?

50. Tom and three friends want to share 39.2 kilograms of squash equally. How much will each person get?

Check your answers on page 204–205.

DECIMALS REVIEW RECORD

Write the number of problems you answered correctly: _____

Write the number of problems you answered incorrectly: _____

Use the Lesson Guide that follows to find out which lessons you should review, if any.

This Lesson Guide can help you make a plan for reviewing decimals. For each problem in the Decimals Review, the lesson that teaches about that kind of problem is listed.

DECIMALS REVIEW LESSON GUIDE

Problem Number	1	2	3	4	5	6	7	8	9	10	11	12
Lesson Number	2	2	2	2	5	5	5	5	12	12	12	12
Problem Number	13	14	15	16	17	18	19	20	21	22	23	24
Lesson Number	21	22	22	23	3	3	3	4	4	14	14	14
Problem Number	25	26	27	28	29	30	31	32	33	34	35	36
Lesson Number	7	7	7	8	9	9	9	9	11	11	11	11
Problem Number	37	38	39	40	41	42	43	44	45	46	47	48
Lesson Number	15	15	15	15	17	19	20	27	29	30	31	15
Problem Number	49	50										
Lesson Number	27	27										

184 *Decimals*

ANSWERS

PART A: WHOLE NUMBERS

Whole Numbers Pretest (page 2)

ADDITION

1. 958
2. 3,799
3. 165
4. 1,459
5. 21,015
6. 225,980
7. 306
8. 545
9. 16,257
10. 15,587

SUBTRACTION

11. 211
12. 450
13. 66
14. 68
15. 3,329
16. 2,588
17. 3,727
18. 50,341
19. 12,205
20. 730,288

MULTIPLICATION

21. 86
22. 1,710
23. 14,436
24. 62,800
25. 281,442
26. 62,775
27. 387,576
28. 294,550
29. 62,118
30. 706,000

DIVISION

31. 28
32. 216
33. 604
34. 920
35. 18
36. 23
37. 36
38. 85
39. 9
40. 13

Fill in the Whole Numbers Pretest Record on page 3.

Chapter 1: ADDITION

Lesson 1 (page 6)

1. 14 60 45
2. 685 450 700
3. 3000 1630 1988
4. 28⑦ 1⑥ 134⓪ 7⑦ 48⑨ 6①
5. 3④5 ①8 25⑧0 4①8 ②6 14⑨2
6. 1②50 ⑨23 7⑦89 2③44 ⑧00 1⑥20
7. ④500 ①878 ①925 ⑥650 ②399 ③776
8. units 2
9. tens 10
10. hundreds 800
11. thousands 3000
12. tens 50
13. thousands 6000

Lesson 2 (page 7)

1. 8 12 8 10 17 12 6 9 11 11
2. 8 8 13 11 9 16 6 15 6 9
3. 13 7 9 4 11 16 10 3 15 7
4. 12 12 10 8 14 7 11 5 12 5
5. 16 7 15 3 8 14 12 2 10 9
6. 9 9 7 10 15 1 3 11 13 6
7. 5 4 17 4 10 13 4 8 13 14
8. 10 15 11 9 9 5 18 10 9 14
9. 6 11 8 7 7 10 6 12 14 7

Lesson 3

EXERCISE A (page 11)

1. 68 79 97 79 95 79
2. 98 49 97 89 86 98
3. 988 877 733 694 928
4. 678 478 699 799 899
5. 6,399 6,977 6,896 9,887
6. 84,567 73,969 94,995 76,993

EXERCISE B (page 12—13)

1. 95 76 97 98
2. 83 69 88 97
3. 465 76 999 189
4. 158 598 868 259
5. 6,899 9,955 7,675
6. 4,547 4,858 48,739
7. 15,669 167,589 39,667

Lesson 4

EXERCISE A (page 15)

1. 111 155 111 101 141 103
2. 120 56 143 140 141 150
3. 760 1050 1190 675 651
4. 1010 513 826 310 723
5. 3,910 9,092 10,993 9,922
6. 38,231 103,521 97,791 119,820

EXERCISE B (page 16)

1. 331 185 512
2. 1363 3993 5216
3. 3303 8200 3340
4. 463 1113 3724
5. 2267 7712 3021

Answers 185

Exercise C (page 17—18)

1. 599   1629
2. 2156   8417
3. 10,909   14,429
4. 36,108   5090
5. 116,691   110,783
6. 158,628   18,054

6. 3,000,000   13,000,000   2,000,000
   40,000,000
7. 17,000,000
8. 20,000,000

Lesson 5 (page 20)

1. |AC| |8| |+| |7| |=|
2. |AC| |3| |7| |5| |+| |6| |6| |=|
3. |AC| |1| |3| |+| |2| |8| |+| |6| |4| |=|
4. |AC| |2| |7| |+| |3| |+| |1| |4| |+| |8| |=|
5. |AC| |2| |0| |6| |3| |+| |4| |9| |=|
6. |AC| |9| |+| |2| |1| |8| |+| |3| |6| |=|

Lesson 6 (page 21)

1. 21 22 15 21 25 13
2. 22 31 21 19 24 30
3. 124 107 101 186 268 295
4. 343 234 244 252 289 288
5. 219 265 195 203 209 248

Lesson 7 (pages 23-24)

1. 230   2. 4,900   3. 6,500,000
4. **b.** five thousand, six hundred
5. **c.** twenty-eight thousand, three hundred
6. **d.** seven million, one hundred thousand
7. two **thousand**, three **hundred**
8. one hundred fifty **thousand**
9. four **million**, eight hundred **thousand**
10. **seven** thousand, **nine** hundred
11. four **hundred** fifteen **thousand**
12. **six** million, seven hundred **thousand**
13. one **thousand**, two **hundred** fifty
14. five **thousand**, two **hundred** eighty
15. seven **million**
16. two hundred thirty-nine **million**
17. fifty-eight **million**
18. 860
19. 4,100
20. 9,650
21. 65,000
22. 2,000,000
23. 180,000
24. 712,000,000
25. 125,760

Lesson 8 (page 26)

1. 540   3,820   14,310   9,050
2. 800   900   26,500   187,600
3. 2,000   19,000   330,000   1,983,000
4. 30,000   20,000   70,000   5,280,000
5. 500,000   600,000   3,100,000   18,500,000

Lesson 9 (pages 27-29)

1. **a.** more
   b. $15,336
        +7,328
        $22,664
2. **a.** sum
   b. $1.99
        +1.49
        $3.48
3. **a.** combined
   b. $2,680
   c. $23,750
        12,480
        +4,200
        $40,430
4. **a.** total
   b. two
   c. 19 ounces and 64 ounces
   d. $2.89
        +2.49
        $5.38
5. **a.** altogether
   b. two
   c. $2.09
        +5.27
        $7.36
6. **a.** total
   b. five
   c. $2.09
        5.27
        0.85
        6.29
        +21.06
        $35.56
7. **a.** combined
   b. four
   c. 129
        183
        87
        +116
        515   pounds
8. **a.** altogether
   b. three
   c. 235
        42
        +36
        313   pounds
9. **a.** total
   b. 515
        2250
        +313
        3078   pounds

Lesson 10 (pages 30-31)

1. $ 6.00
   +9.00
   $15.00
2. $6.00
   2.00
   +1.25
   $9.25
3. $7.00
4. $6.00
   +1.50
   $7.50

5. $ 6.00
   +9.00
   $15.00

6. $ 6.00
     1.50
   +23.85
   $31.35

Lesson 11 (pages 32-33)

1. $1.95
   +0.65
   $2.60

2. $1.25
   +0.50
   $1.75

3. $1.35
   0.50
   1.35
   +0.50
   $3.70

4. $1.95
   2.95
   0.50
   +0.75
   $6.15

5. $0.50
   +1.95
   $2.45

6. $ 1.95
     3.50
     0.65
     1.95
     1.95
     0.50
     1.35
   +1.80
   $13.65

Lesson 12

Exercise A (page 35)

1. Car Sales at Al's Autos
2. cars sold
3. 0   4. 100   5. four   6. 60 cars
7. 90 cars   8. 80 cars   9. July
10. May   11. 40
              60
              90
           +80
           270 cars
12. 80
    +35
    115 cars

Exercise B (page 36)

1. Weight of Larcom Family Members
2. pounds   3. 0   4. 200   5. four
6. 70 pounds   7. 110 pounds
8. 180 pounds   9. 60
                 +70
                 130 pounds
10. 110
    +18
    128 pounds

11. 180
    +22
    202 pounds

Chapter 2: SUBTRACTION

Lesson 13 (pages 37-38)

1. 7 1 6 7 4 3 4 9 5 3
2. 5 3 4 5 8 0 7 6 4 2
3. 4 5 1 0 7 5 1 7 2 0
4. 9 4 8 2 9 8 1 5 7 9
5. 6 2 6 3 7 9 1 2 8 6
6. 8 9 1 6 9 7 1 4 2 3
7. 4 4 7 8 0 2 5 5 8 3
8. 3 9 6 8 7 0 4 6 3 2
9. 8 3 9 0 6 5 9 1 2 5

Lesson 14 (page 40)

1. 411   32   252
2. 352   23   318
3. 2100   1002   3341
4. 7213   6701   310
5. 11,162   52,011   84,201
6. 22,114   51,104   12,115

Lesson 15

Exercise A (page 41)

1. 19   28   9   69
2. 38   39   69   17
3. 706   109   205

Exercise B (page 42)

1. 199   152   39
2. 165   237   489
3. 8612   5508   7082
4. 928   4702   2132
5. 22,591   69,059   10,918
6. 26,167   20,863   64,769
7. 289,150   156,861   281,098
8. 5,766   249,065   84,389
9. 505,060   808,000

Lesson 16

Exercise A (page 44)

1. 51   8   32
2. 359   59   65
3. 511   108   21

Answers   187

EXERCISE B (pages 45-46)

1. 268 693 247
2. 212 373 364
3. 865 50 759
4. 1595 3,334 1960
5. 529 1,651 21,059
6. 18,694 41,481 65,945
7. 139,470 60,875
8. 1,285,000 8,143,600
9. 5,144,800 4,147,000

Lesson 17 (page 47)

1. 32 412 544
2. 562 133 136
3. 431 531 408
4. 865 5791 3837
5. 1143 3703 1517
6. 816 511 4511
7. 1823 3809 5650
8. 8170 607 4115
9. 10,885 38,621 17,117
10. 7,493 22,186 35,114
11. 110,912 203,882
12. 600,438 218,744
13. 603,525 3,909,000

Lesson 18 (pages 48-50)

1. a. how much more
   b. $26,215
      −25,977
      $ 238
2. a. how much less
   b. 87 pounds
   c. 183
      −129
        54 pounds
3. a. Texas and Florida
   b. $59.95
   c. $79.95
      −44.95
      $35.00
4. a. 1988
   b. 1988
      −1896
          92
5. a. 1988
   b. 1988
      −1891
          97
6. a. 30 inches
   b. $169
      −129
      $ 40
7. a. 226,545,805
   b. 75,994,575
      −5,308,483
      70,686,092
8. a. 5,308,483
   b. 226,545,805
      −75,994,575
      150,551,230

9. a. how many more
   b. 80,089
      −66,074
      14,015
10. a. how many more
    b. 34,921,000
       −28,609,000
        6,312,000

Lesson 19 (pages 51-53)

1. a. $275 rent & $430 food
   b. add
   c. subtract
   d. $275 $1180
      +430   − 705
      $705 $ 475
2.   8   25
    +11   −19
     19   6 chairs
3. 32   50
    9   −47
   +6   3 pounds
   47
4. a. more than $329
   b. add
   c. $329
      + 50
      $379
5. $13,600
   + 2,000
   $15,600
6. a. subtract
   b. subtract
   c. add
   d. $37.95 $47.90 $30.45
      − 7.50   −10.99   +36.91
      $30.45 $36.91 $67.36
7. a. two
   b. $14.95
      +19.50
      $34.45
8. a. add
   b. subtract
   c. $34.45 $40.00
      + 2.06   −36.51
      $36.51 $ 3.49
9. a. three
   b. $14.95
       18.00
      +19.50
      $52.45
10. $52.45 $60.00
    + 3.20   −55.65
    $55.65 $ 4.35

Lesson 20 (page 54)

Item Number	Date	CHECK REGISTER Description	Subtractions		Additions		BALANCE	
							125	13
62	6/23	County Savings Bank	72	88			−72	88
							52	25
—	6/25	deposit			436	17	+436	17
							488	42
—	6/28	service charge	6	00			−6	00
							482	42
63	7/1	Munro Management	265	00			−265	00
							217	42
64	7/7	General Telephone	24	79			−24	79
							192	63
—	7/8	deposit			436	17	+436	17
							628	80
65	7/14	Edison Utilities	26	54			−26	54
							602	26
66	7/16	Mel's Market	128	76			−128	76
							473	50

Lesson 21 (page 55)

1. 252 2. 177 3. 47 4. 45

5. 252   6. 258   7. 182
   +182   +177   −177
    434   435   5

8. 1987 9. 53   10. 258
           −47   −177
             6   81

Lesson 22 (pages 56-57)

1. Springfield Hawks Wins and Losses
2. games   3. wins
4. losses 5. three
6. 26 29 22   7. 6 7 8
8. 26   9.   8
    −6   −7
    20   1
10. a. 26   b. 29   c. 22
       +6   +7   +8
        32   36   30
11. 1986   12. 1985

Chapter 3: MULTIPLICATION

Lesson 23 (pages 58-59)

1. 27 7 60 55 28 0 16 48 18 63
2. 56 24 16 60 72 6 80 8 121 30
3. 44 18 96 77 36 81 35 8 24 0
4. 40 144 12 90 48 33 20 20 84 60
5. 72 15 0 49 32 12 30 110 80 21
6. 10 54 40 108 88 9 18 36 36 55
7. 25 110 14 33 45 5 18 56 0 42
8. 84 0 30 9 72 66 90 1 24 12
9. 21 4 44 16 14 20 48 48 99 28
10. 32 12 36 40 54 6 10 0 96 40
11. 60 66 45 42 0 36 6 100 27 7
12. 8 63 24 0 64 35 77 108 15 72

Lesson 24 (pages 61-62)

1. 129   306   300   568
2. 427   168   455   287
3. 729   420   168   96
4. 248   568   400   637
5. 846   3066   693   5608
6. 3570   2484   7209   1486

Lesson 25 (pages 63-64)

1. 168   292   336   195
2. 57   100   196   342
3. 415   603   150   486
4. 236   92   54   368
5. 798   2070   3105   3252
6. 6328   1442   1494   5094
7. 36,832   19,432   39,204   9,460

Lesson 26 (pages 65-67)

1. 1848   748   3168
2. 3286   7980   2442
3. 1240   2320   6160
4. 1036   7440   2912
5. 2646   1710   3864
6. 592   1512   2697
7. 2208   7695   4818
8. 7597   14,758   14,272
9. 9810   45,920   6741
10. 31,160   23,092   47,430
11. 39,376   41,099   13,620
12. 63,360   102,028   46,431
13. 55,315   241,135   379,660

Lesson 27 (pages 68-69)

1. 325,413   366,240   371,332
2. 58,646   592,749   404,320
3. 588,528   312,417   126,420
4. 215,400   493,500   338,400
5. 808,562   3,628,644   1,425,476
6. 4,772,752   4,249,920   2,212,074

Lesson 28 (page 70)

1. 370   3,600   1,450   18,000
2. 8,300   20,000   24,800   990
3. 16,000   4,000   2,500   39,200
4. 10,000   140,000   8,500   200,000
5. 8,000   3,200   410,000   2,500
6. 720 pounds
7. $3,500
8. $2,478,000

Lesson 29 (pages 71-73)

1. a. 21 miles per gallon
   b. 21
      ×13
       63
       21
      273   miles

2. a. 36 pounds per package
   b. 36
      ×19
      324
       36
      684   pounds

3. a. $2486 per student
   b. $2,486
       ×30
      $74,580

4. $2,486 × 1000 = $2,486,000

5. a. $3.35 per hour
   b. $5.00
   c. $3.35
      × 40
      $134.00

6. a. $18 per month
   b. $18
      ×12
       36
       18
      $216

7. a. $7.95 per room
   b. $7.95
      ×   5
      $39.75

8. a. 67 words per minute
   b. 67
      ×30
      2010   words

9. a. $1.49 per pound
   b. 12 pounds
   c. $1.49
      × 12
       2 98
      14 9
      $17.88

10. a. $1.29 per pound
    b. 10 pounds
    c. $1.29
       × 10
       $12.90

Lesson 30 (pages 74-75)

1. a. $9   b. $12   c. $72
      ×8   ×3   ×36
      $72 $36 $108

2. a. $24.99   b. $74.97
       × 3   −59.99
      $74.97 $14.98

3. a. $19   b. add   c. $228
      ×12   + 75
       38               $303
       19
      $228

4. a. 525   b. 175   c. $3150
       × $6   × $10   + 1750
       $3150 $1750 $4900

5. a. 18   b. 27
       × 8   × 6
       144 pounds   162 pounds

   c. 144
       + 162
       306 pounds

6. a. $25   b. subtract   c. $1000
       × 27   − 675
       175                        $ 325
       50
       $675

7. a. 83   b. 3   c. 218
       72   × 3
       + 63   654 pounds
       218 pounds

Lesson 31 (pages 76-77)

1. $D = RT$ 2. $D = RT$
   $D = 63 \times 4$ $D = 4 \times 17$
   $D = 252$ miles $D = 68$ miles

3. $D = RT$ 4. $D = RT$
   $D = 3 \times 4$ $D = 535 \times 6$
   $D = 12$ miles $D = 3210$ miles

5. $D = RT$
   $D = 79 \times 3$
   $D = 237$ miles

6. $D = RT$ $D = RT$ 22
   $D = 11 \times 2$ $D = 56 \times 3$ + 168
   $D = 22$ miles $D = 168$ miles   190 miles

7. $D = RT$ $D = RT$
   $D = 9 \times 2$ $D = 17 \times 2$
   $D = 18$ miles $D = 34$ miles

   $D = RT$ 18
   $D = 63 \times 4$ 34
   $D = 252$ miles   + 252
                      304 miles

Lesson 32 (pages 78-79)

1. $P = 2L + 2W$
   $P = 2 \times 15 + 2 \times 6$
   $P = 30 + 12$
   $P = 42$ yards

2. $P = 2L + 2W$
   $P = 2 \times 20 + 2 \times 8$
   $P = 40 + 16$
   $P = 56$ meters

3. $P = 2L + 2W$
   $P = 2 \times 63 + 2 \times 42$
   $P = 126 + 84$
   $P = 210$ inches

4. $P = 2L + 2W$
   $P = 2 \times 18 + 2 \times 9$
   $P = 36 + 18$
   $P = 54$ feet

5. $P = 2L + 2W$
   $P = 2 \times 25 + 2 \times 12$
   $P = 50 + 24$
   $P = 74$ feet

   Notice the word *perimeter* is not in this problem.

6. $P = 2L + 2W$
   $P = 2 \times 17 + 2 \times 11$
   $P = 34 + 22$
   $P = 56$ inches

7. $P = 2L + 2W$
   $P = 2 \times 5 + 2 \times 3$
   $P = 10 + 6$
   $P = 16$ feet

Lesson 33 (pages 80-81)

1. $A = LW$
   $A = 18 \times 10$
   $A = 180$ square feet

2. $A = LW$
   $A = 16 \times 11$
   $A = 176$ square inches

3. $A = LW$
   $A = 14 \times 7$
   $A = 98$ square yards

4. $A = LW$
   $A = 19 \times 5$
   $A = 95$ square yards

5. $A = LW$
   $A = 5 \times 4$
   $A = 20$ square yards

6. $20 $400
    × 20   + 35
    $400 $435

7. $A = LW$
   $A = 18 \times 12$
   $A = 216$ square feet

8. $A = LW$
   $A = 12 \times 9$
   $A = 108$ square feet

9.
```
   216        324
  +108       ×  $5
  ───       ─────
   324       $1620
square feet
```

Lesson 34 (pages 82-83)

1. $V = LWH$
   $V = 13 \times 9 \times 8$
   $V = 936$ cubic feet

2. $V = LWH$
   $V = 20 \times 15 \times 11$
   $V = 3300$ cubic inches

3. $V = LWH$
   $V = 18 \times 10 \times 8$
   $V = 1440$ cubic feet

4. $V = LWH$
   $V = 14 \times 6 \times 1$
   $V = 84$ cubic yards

5. $V = LWH$
   $V = 30 \times 15 \times 5$
   $V = 2250$ cubic feet

6. $V = LWH$
   $V = 6 \times 5 \times 60$
   $V = 1800$ cubic feet

Lesson 35 (pages 84-85)

1. Minnesota
2. Alabama
3. $99 × 4 = $396
4. $156 × 4 = $624
5. $160 × 15 = $2400 (800 + 160)
6. $118 × 23 = $2714 (354 + 236)
7. $86 × 4 = $344
8. $344 + $278 = $622
9. $135 × 4 = $540
10. No

11. $622 − 540 = $82 needed

Lesson 36 (pages 86-87)

1. a. $97   b. $98   c. $105   d. $112
2. $98 × 4 = $392
3. $112 × 52 = $5824 (224 + 560)
4. $97 × 4 = $388; $105 × 4 = $420; $420 − 388 = $32   or $105 − 97 = $8; $8 × 4 = $32
5. $112 × 4 = $448; $1297 − 448 = $849

Chapter 4:   DIVISION

Lesson 37 (page 88)

1. 7 7 2 0 6 1 9 3
2. 4 3 6 5 5 2 7 3
3. 2 3 6 0 8 7 9 3
4. 9 8 2 1 4 9 7 8
5. 8 4 7 2 8 5 6 9
6. 5 1 0 8 4 7 3 0
7. 5 2 8 6 5 9 8 6
8. 9 5 4 7 0 6 2 9
9. 5 3 4 5 9 7 1 4
10. 1 0 9 6 8 3 7 2

Lesson 38 (pages 91-92)

1. 17   52   73   44
2. 35   81   27   69
3. 32   46   91   74
4. 53   25   87   68
5. 62   48   94   36
6. 312   743   126   522
7. 871   426   238   692
8. 653   924   236   144
9. 329   473   812   726
10. 5,133   4,327   7,284
11. 9,115   8,518   7,362
12. 6,392   4,436   7,245

Lesson 39 (page 93)

1. 25 r 2   52 r 5   47 r 1   49
2. 18 r 3   38 r 1   63   71 r 3
3. 86   92 r 4   48 r 2   59 r 1
4. 512   934 r 2   346 r 8   615
5. 722 r 6   418   236 r 2   847 r 1
6. 1314 r 4   2,763   8,522 r 7   6,731

Lesson 40 (page 94)

1. 70   40 r 7   30 r 8   90 r 3
2. 42   30 r 3   80   60 r 5
3. 204   373   510 r 2   408
4. 340   506   309   431 r 4
5. 210   723 r 2   807   690

Lesson 41 (pages 95-96)

1. 4   7   3
2. 6   2 r 12   5
3. 7 r 30   9   4 r 18
4. 8   3 r 23   6
5. 12 r 4   18   21
6. 32 r 26   26   51
7. 24   40 r 21   63
8. 29 r 5   81   56
9. 67   72   30 r 8
10. 413   528   334
11. 306 r 15   480   116
12. 274   207 r 20   526

Lesson 42 (pages 97-98)

1. 3   4 r 28   6
2. 5   7 r 195   8
3. 9   2   4 r 320
4. 20 r 180   32   16
5. 41   33 r 180   56

Lesson 43 (pages 99-100)

1. 9   7   6
2. 5   43   85
3. 12   195   230
4. 21   50   700
5. 165   48   365
6. 48   600   120
7. $8
8. 123 pounds
9. $680

Lesson 44 (pages 101-103)

1. a. dollars
   b. $6240
   c. $\quad$ $ 480
      $\overline{13)\$6240}$
      $\underline{52}$
      104
      $\underline{104}$
      00

2. a. cubic feet
   b. 222
   c. three
   d. $\quad$ 74 cubic
      $\overline{3)222}$ feet
      $\underline{21}$
      12
      $\underline{12}$
      0

3. a. pounds
   b. 92 pounds
   c. $\quad$ 23 pounds
      $\overline{4)92}$
      $\underline{8}$
      12
      $\underline{12}$
      0

4. a. dollars
   b. $315
   c. $\quad$ $ 9
      $\overline{35)\$315}$
      $\underline{315}$
      0

5. a. miles
   b. 391 miles
   c. $1.09
   d. $\quad$ 23 miles
      $\overline{17)391}$
      $\underline{34}$
      51
      $\underline{51}$
      0

6. a. dollars
   b. $1800
   c. $1800 ÷ 10
      = $180

7. a. dollars
   b. $126,000
   c. $\quad$ $ 1,050
      $\overline{120)\$126,000}$
      $\underline{120}$
      6 0
      $\underline{0}$
      6 00
      $\underline{6\ 00}$
      00

8. a. ounces
   b. 432 ounces
   c. $\quad$ 12 ounces
      $\overline{36)432}$
      $\underline{36}$
      72
      $\underline{72}$
      0

Lesson 45 (pages 104-105)

1. a. miles
   b. division
   c. $\quad$ 24 miles
      $\overline{23)552}$
      $\underline{46}$
      92
      $\underline{92}$
      0

2. a. $1.09
   b. multiplication
   c. $\quad$ $1.09
      $\quad\times\quad$ 23
      $\quad$ 3 27
      $\quad$ 21 8
      $\quad$ $25.07

3. a. less
   b. division
   c. 33 rows
      $19\overline{)627}$
      $\underline{57}$
      57
      $\underline{57}$
      0

5. a. 113
   b. multiplication
   c. 113
      $\underline{\times56}$
      678
      $\underline{565}$
      6328 bushels

7. a. division
   b. four
   c. $\$4{,}746$
      $4\overline{)\$18{,}984}$
      $\underline{16}$
      29
      $\underline{28}$
      18
      $\underline{16}$
      24
      $\underline{24}$
      0

4. a. $4
   b. multiplication
   c. 627
      $\underline{\times\$4}$
      $\$2508$

6. a. $3
   b. multiplication
   c. $6{,}328$
      $\underline{\times\$3}$
      $\$18{,}984$

6. 1119 diners
   $203\overline{)57}$
   $\underline{+26}$
   57

7. 6276 televisions
   $854\overline{)304}$
   101
   $\underline{+56}$
   304

Lesson 47 (page 108)

2. Sausage $\$.29$
   $8\overline{)\$2.32}$
   $\underline{16}$
   72
   $\underline{72}$
   0

3. Bologna $\$.15$
   $16\overline{)\$2.40}$
   $\underline{16}$
   80
   $\underline{80}$
   0

4. Ham $\$.27$
   $9\overline{)\$2.43}$
   $\underline{18}$
   63
   $\underline{63}$
   0

5. White bread $\$.05$
   $24\overline{)\$1.20}$
   $\underline{120}$
   0

6. Cheddar cheese $\$.20$
   $16\overline{)\$3.20}$
   $\underline{32}$
   00

7. Lunch meats $\$.11$
   $8\overline{)\$.88}$
   $\underline{8}$
   08
   $\underline{8}$
   0

8. Apple sauce $\$.03$
   $33\overline{)\$.99}$
   $\underline{99}$
   0

Lesson 46 (pages 106-107)

1. 8581
   $694\overline{)324}$
   93
   $\underline{+77}$
   324

2. 1928
   $383\overline{)84}$
   $\underline{+27}$
   84

3. 912 pounds
   $\underline{+15}2\overline{)24}$
   24

4. $\$85{,}000\$82{,}000$
   $72{,}0003\overline{)\$246{,}000}$
   $\underline{+89{,}000}$
   $\$246{,}000$

5. $\$14.20\$27.46$
   $39.753\overline{)\$82.38}$
   $\underline{+28.43}$
   $\$82.38$

194 Answers

Lesson 48

Exercise A (page 109)

1.
50	18	10°C		95	63	35°C
−32	×5	9)90		−32	×5	9)315
18	90			63	315	

68	36	20°C		59	27	15°C
−32	×5	9)180		−32	×5	9)135
36	180			27	135	

104	72	40°C
−32	×5	9)360
72	360	

2.
41	9	5°C		86	54	30°C
−32	×5	9)45		−32	×5	9)270
9	45			54	270	

140	108	60°C		122	90	50°C
−32	×5	9)540		−32	×5	9)450
108	540			90	450	

212	180	100°C
−32	×5	9)900
180	900	

Exercise B (page 110)

1.
200	360	360		80	144	144
×9	5)1800	+32		×9	5)720	+32
1800		392°F		720		176°F

30	54	54		75	135	135
×9	5)270	+32		×9	5)675	+32
270		86°F		675		167°F

5	9	9
×9	5)45	+32
45		41°F

2.
95	171	171		60	108	108
×9	5)855	+32		×9	5)540	+32
855		203°F		540		140°F

100	180	180		150	270	270
×9	5)900	+32		×9	5)1350	+32
900		212°F		1350		302°F

10	18	18
×9	5)90	+32
90		50°F

Whole Numbers Review (pages 111–115)

1. 5,989
2. 1,415
3. 7,585
4. 554,947
5. 218
6. 3,068
7. 8,026
8. 112,512
9. 3,408
10. 18,441
11. 168,294
12. 125,132
13. 473
14. 308
15. 15
16. 63
17. 470
18. 71,000
19. 306,000
20. 1,814,000
21. 190
22. 2,400
23. 11,000
24. 130,000
25. $1.59 + 3.29 = $4.88
26. $472 + 468 + 234 = $1174
27. 3,655,000 + 3,634,000 = 7,289,000
28. $19,270 + 1,560 = $20,830
29. $264.98 − 229.97 = $35.01
30. 1988 − 1776 = 212
31. $6.99 + 8.99 + .72 = $16.70; $20.00 − 16.70 = $3.30
32. $393 − 350 = $43
33. $684 − 385 = $299
34. 6 + 2 + 30 + 12 = 50 gallons
35. 30 − 20 = 10 gallons
36. $68,500 − 9,000 = $59,500
37. $15.95 × 24 = 63 80 / 319 0 / $382.80
38. 17 × 30 = 510 pounds
39. $32.50 × 4 = $130.00; $130.00 + 8.45 = $138.45; $150.00 − 138.45 = $11.55
40. 59 × 3 = 177 miles
41. $P = 2L + 2W$
    $P = 2 \times 18 + 2 \times 13$
    $P = 36 + 26$
    $P = 62$ feet

42. $A = LW$
    $A = 18 \times 13$
    $A = 234$ square feet

43.
```
  $45      $1620
  ×36     +1200
  ---     -----
  270     $2820
  135
  -----
 $1620
```

44.
```
   $ 2.65
6)$15.90
```

45.
```
    $319
3)$957
```

46.
```
    19 miles
26)494
   26
   ---
   234
   234
   ---
     0
```

47.
```
     $ 1.07
26)$27.82
    26
    ---
     1 8
       0
     ---
     1 82
     1 82
     ----
        0
```

48.
```
     38 mph
13)494
   39
   ---
   104
   104
   ---
     0
```

49.
```
    6         7 hours
   10       3)21
   +5
   ---
   21
```

50. $1800 \div 100 = 18$ pounds

Part B: Decimals

Decimals Pretest (pages 118-119)

Addition and Subtraction

1. 1.6
2. .779
3. 2.258
4. 11.82
5. .0873
6. 29.966
7. .3
8. .54
9. 2.63
10. 1.905
11. .095
12. 6.722

Multiplication

13. .72
14. .3
15. .008
16. 9.52
17. 19.26
18. 5.43
19. .0648
20. .00126
21. 260
22. 6200

Division

23. 1.3
24. .24
25. 3.2
26. .28
27. 9
28. 70
29. 40
30. 1300
31. .262
32. .0048

Fill in the Decimals Pretest Record on page 119.

Chapter 1: ADDITION AND SUBTRACTION

Lesson 1 (pages 120-121)

1. 10.3 .4 7.4 .8
2. 23.56 .75 1.17
3. 6.053 .488 1.675
4. 4.⑤ .⑥ 20.①5 2.⑥78 .③5 2.⑧
5. .0⑤ .2④ 2.1⑥ .3⑧75 2.8⑦ .3⑥7
6. .12⑤ 2.60⑧7 .00⑨3 1.00③ .74⑨ 25.12⑧
7. units 3
8. tenths 10
9. hundredths 100
10. thousandths 1000

Lesson 2 (pages 123-125)

1. b. 2.5
       +.36
       2.86

2. a. .8
       .43
       +.128
       1.358

3. b. 9.
       +2.37
       11.37

4. c. .03
       .4
       +.069
       .499

5. b. 1.3
       .88
       +4.
       6.18

6. 1.1 1.26 1.145
7. 6.889 93.27 423.9
8. 36.14 22.059 1.3933
9. 16.75 10.886 35.742
10. 22.484 853.636
11. 15.535 1.315
12. 13.527 3.472

Lesson 3

Exercise A (page 126)

1. .2
2. .08
3. .015
4. .27
5. .219
6. .0026
7. five **tenths** six tenths
8. nine **hundredths** thirteen hundredths
9. twelve **hundredths** four thousandths
10. twenty-five **thousandths** nineteen thousandths

Exercise B (pages 127-128)

1. 4.9
2. 80.02
3. 3.015
4. four and one tenth
5. eight and six tenths
6. two and sixteen hundredths
7. twenty and nineteen hundredths
8. seventeen and three thousandths
9. five and fourteen thousandths
10. two and two tenths
11. two and fifty-four hundredths
12. one and six tenths
13. .3
14. .13
15. .004
16. .0005
17. .16
18. 12.6
19. 5.009
20. 60.8
21. 3.011
22. 14.06

Lesson 4 (page 129)

1. .83 .6 2.55
2. 3.4 .04 .047
3. 10.2 .106 .2
4. .717 .3
5. .11 1.202
6. 3.88 .413
7. .12 pound
8. 1.1 meters

Answers 197

Lesson 5 (page 131)

1. .09 .225 .44
2. 1.32 8.065 .35
3. .083 .507 10.43
4. 2.724 5.66 1.72
5. .504 4.62 .13
6. 1.6 .081 15.6
7. .025 15.656 .588

Lesson 6 (pages 132-133)

1. 10.45 .55
2. 9.34 .017
3. 71.04 12.306
4. 5.58 .0063
5. 353.68 1.726
6. 2.565 44.339
7. 8.242 .017
8. 23.53
9. 5.512
10. .077
11. 27.35
12. .109
13. 13.96
14. 5.49
15. .376
16. 28.79
17. 1.213
18. .041
19. .6626
20. 17.1
21. 4.433
22. 3.267

Lesson 7 (pages 134-135)

1. a. total
   b. 8.5
       5.2
       6.7
     +12.3
       32.7   miles

2. a. how much more
   b. $148.0
      −142.6
       $ 5.4   billion

3. a. how much greater
   b. 170.0
      −116.1
       53.9 million

4. a. how much greater
   b. 116.1
      −101.8
       14.3 million

5. a. difference
   b. $1,460.0
      − 666.8
       $ 793.2 million

6. a. total
   b. 6.3
       4.25
     +3.95
       14.50 = 14.5 pounds

7. a. how much more
   b. 178.1
      −108.5
       69.6 pounds

Lesson 8 (pages 136-137)

1. a. addition
   b. subtraction
   c. 101.4   103.2
     + 1.8   − 4.3
      103.2°   98.9°

2. a. $2 million
   b. subtraction
   c. .35
       1.2
      + .2
       1.75

      $2.00
     −1.75
      $ .25   million

3. a. add
   b. subtract
   c. 8.5
       6.25
       9
       11.5
      +9.25
       44.50

       44.5
      −40.0
        4.5   hours

4. a. subtract
   b. add
   c. 9133.2
      −8916.4
        216.8

        216.8
       + 85
        301.8   miles

Lesson 9 (page 138)

1. a. 5.4 pounds
   b. 4.7 pounds
   c. 9.3 pounds
   d. 10.8 pounds

2. 1970 and 1985

3. 1980

4. 1960

5. 7.7
   −5.4
    2.3   pounds

6. 11.4
   − 4.5
    6.9   pounds

7. 11.4
   −10.8
     .6   pound

198   Answers

Lesson 10 (pages 139-140)

1. Snowfall in Capital City
2. inches 3. four 4. January
5. November 6. 4 and 5 7. 4.5 inches
8. one 9. 4.1 inches 10. 2.8 inches
11. 3.4 inches 12. 4.5
    -2.8
    1.7 inches

Lesson 11 (page 142)

1. a. *A* 1.5 cm
   b. *B* 2.3 cm
   c. *C* 3.6 cm
   d. *D* 4.1 cm
   e. *E* 7.8 cm
   f. *F* 9.2 cm

2. a. *A* and *B* 2.3
                  −1.5
                   .8 cm
   b. *A* and *C* 3.6
                  −1.5
                   2.1 cm
   c. *B* and *D* 4.1
                  −2.3
                   1.8 cm
   d. *C* and *E* 7.8
                  −3.6
                   4.2 cm
   e. *C* and *F* 9.2
                  −3.6
                   5.6 cm

Chapter 2: MULTIPLICATION

Lesson 12

Exercise A (page 144)

1. .12   5.28   .884
2. 49.12   2.848   345.6
3. 149.6   4.307   32.2

Exercise B (page 145)

1. .072   .0368   .0045
2. .084   .00216   .072

Exercise C (page 146)

1. .2   27   7.6
2. 1.17   5.4   105
3. 4.42   55.8   3.87
4. .16   3.3   .378
5. 69.3   .21   23.8
6. 46.4   19.44   .033
7. 1.512   .114   17.3
8. .0153   .073   6.24
9. 27   .008   .129
10. 2.55   86.4   247

Lesson 13 (page 147)

1. 140   6800   53
2. 300   4.7   70
3. 2.8   36.5   180
4. 1640   8700   42.5
5. 1280   73.33   2746
6. 52   130   30,500
7. 120 pounds
8. 8 miles

Lesson 14 (pages 148-149)

1. 1.4   27.3   .5 $5.10 $8.40
2. 1.39   .30   1.39 $2.88 $.31
3. .225   8.424   .333   10.597   2.886
4. 3   17   8 $6 $20
5. 6.6 inches
6. 7 inches

Lesson 15 (pages 150-151)

1. a. 6.5 hours per week
   b.   6.5
      ×1 6
       39 0
       65
      104.0 = 104 hours

2. a. 1.3 ounces
   b.   1.3
      ×1 5
       6 5
       13
      19.5 ounces

3. a. $5.35
   b. $5.3 5
       × 3 7.5
         2 6 7 5
        37 4 5
       160 5
       $200.6 2 5 to the nearest penny
       = $200.63

4. a. 36 mph
   b.   3 6
      ×1.5
       180
       36
       54.0
       = 54 miles

5. a. $1.089
   b. $1.089
      ×   13
        3 267
       10 89
       $14.157 to the nearest penny
       = $14.16

6. a. $.78
   b. $.78
       × 3.25
         3 90
        15 6
        2 34
       $2.53 50 to the nearest penny
       = $2.54

Answers 199

7. a. 2.54 cm
   b. 2.54
      × 12
      ─────
       5 08
      25 4
      ─────
      30.48

Lesson 16 (pages 152-153)

1. a. 4.5
      ×6 0
      ─────
      270.0 miles
   b. 1.5
      ×3 2
      ─────
       3 0
       45
      ─────
      48.0
   c. add
   d. 270
      +48
      ─────
      318 miles

2. a. $1.49
   b. $1.4 9
      ×   3.5
      ─────
       7 4 5
       4 4 7
      ─────
      $5.2 1 5 to the nearest penny = $5.22
   c. subtract
   d. $10.00
      −5.22
      ─────
      $ 4.78

3. a. $5.50
      ×   40
      ─────
      $220.00
   b. $8.2 5
      ×   6.5
      ─────
       4 1 2 5
       49 5 0
      ─────
      $53.6 2 5 to the nearest penny = $53.63
   c. addition
   d. $220.00
      +53.63
      ─────
      $273.63

4. a. $6.8 0
      × 1 2.5
      ─────
       3 4 0 0
       13 6 0
      ─────
      $85.0 0 0
   b. add
   c. $85
      +75
      ─────
      $160

5. a. 24.5
      × 3 0
      ─────
      735.0
   b. $4.60
   c.   7 35
      × $4.60
      ─────
       441 00
       2940
      ─────
      $3381.00

Lesson 17 (page 154)

1. $P = 2L + 2W$
   $P = 2 \times 7 + 2 \times 3.5$
   $P = 14 + 7$
   $P = 21$ feet

   $A = LW$
   $A = 7 \times 3.5$
   $A = 24.5$ square feet

2. $P = 2L + 2W$
   $P = 2 \times 6.6 + 2 \times 4.2$
   $P = 13.2 + 8.4$
   $P = 21.6$ meters

   $A = LW$
   $A = 6.6 \times 4.2$
   $A = 27.72$ square meters

3. $P = 2L + 2W$
   $P = 2 \times 5.4 + 2 \times 3.2$
   $P = 10.8 + 6.4$
   $P = 17.2$ inches

   $A = LW$
   $A = 5.4 \times 3.2$
   $A = 17.28$ square inches

4. $P = 2L + 2W$
   $P = 2 \times 16 + 2 \times 7.5$
   $P = 32 + 15$
   $P = 47$ yards

   $A = LW$
   $A = 16 \times 7.5$
   $A = 120$ square yards

5. $P = 2L + 2W$
   $P = 2 \times 12.5 + 2 \times 2.6$
   $P = 25 + 5.2$
   $P = 30.2$ meters

   $A = LW$
   $A = 12.5 \times 2.6$
   $A = 32.5$ square meters

6. $P = 2L + 2W$
   $P = 2 \times 10 + 2 \times 4.8$
   $P = 20 + 9.6$
   $P = 29.6$ inches

   $A = LW$
   $A = 10 \times 4.8$
   $A = 48$ square inches

Lesson 18

Exercise A (page 156)

1. $d = 2r$
   $d = 2 \times .6$
   $d = 1.2$ inches

2. $d = 2r$
   $d = 2 \times 4.25$
   $d = 8.5$ yards

3. $d = 2r$
   $d = 2 \times 1.8$
   $d = 3.6$ meters

4. $d = 2r$
   $d = 2 \times 10.15$
   $d = 20.3$ feet

Exercise B (page 156)

1. $r = .5d$
   $r = .5 \times 15$
   $r = 7.5$ yards
2. $r = .5d$
   $r = .5 \times 24$
   $r = 12$ feet
3. $r = .5d$
   $r = .5 \times 3.4$
   $r = 1.7$ meters
4. $r = .5d$
   $r = .5 \times .5$
   $r = .25$ inch

Lesson 19 (page 157)

1. $C = \pi d$
   $C = 3.14 \times 8$
   $C = 25.12$ feet
2. $C = \pi d$
   $C = 3.14 \times 15$
   $C = 47.1$ yards
3. $C = \pi d$
   $C = 3.14 \times 10$
   $C = 31.4$ meters
4. $C = \pi d$
   $C = 3.14 \times 43$
   $C = 135.02$ inches
5. $C = \pi d$
   $C = 3.14 \times 50$
   $C = 157$ feet
6. $C = \pi d$
   $C = 3.14 \times 65$
   $C = 204.1$ inches

Lesson 20

Exercise A (page 158)

1. $7^2 = 7 \times 7 = 49$
2. $12^2 = 12 \times 12 = 144$
3. $9^2 = 9 \times 9 = 81$
4. $1.3^2 = 1.3 \times 1.3 = 1.69$

Exercise B (page 159)

1. $A = \pi r^2$
   $A = 3.14 \times 3 \times 3$
   $A = 28.26$ square feet
2. $A = \pi r^2$
   $A = 3.14 \times 20 \times 20$
   $A = 1256$ square inches
3. $A = \pi r^2$
   $A = 3.14 \times 5 \times 5$
   $A = 78.5$ square yards
4. $A = \pi r^2$
   $A = 3.14 \times 8 \times 8$
   $A = 200.96$ square meters
5. $A = \pi r^2$
   $A = 3.14 \times 30 \times 30$
   $A = 2826$ square feet
6. $A = \pi r^2$
   $A = 3.14 \times 4 \times 4$
   $A = 50.24$ square yards

Chapter 3: DIVISION

Lesson 21 (page 161)

1. 2.6   .18   5.8
2. .79   .036   6.3
3. .07   .24   .036
4. .43   .009   8.5
5. 2.07   34.3   .107
6. 8.2   .067   .58

Lesson 22 (page 164)

1. 24   13   36
2. 1.7   4.2   190
3. .22   3.4   .015
4. 7.33   140   1.48
5. 30.8   2.19   4250
6. 210   230   .016
7. 4.3   240   .38
8. 73   54   2.7
9. 5.9   .062   830
10. 32.3   80.4   4.18
11. 63.1   27   5.06
12. 70   .09   .016
13. 4   2.3   180

Lesson 23 (pages 165-166)

1. 40   70   700
2. 500   60   80
3. 3000   200   900
4. 700   8000   4000
5. 160   2700   430
6. 3200   240   810

Lesson 24 (page 167)

1. 60   9   70
2. 5.2   .02   .3
3. 80   .032   .07
4. 24   700   18
5. 5.2   35   430
6. 500   6.3   800
7. 1.3   .035   90
8. .026   6000   .08

Lesson 25 (page 169)

1. a. 7.33 to the nearest tenth = 7.3
   b. 5.67 to the nearest tenth = 5.7
   c. .46 to the nearest tenth = .5
2. a. 26.666 to the nearest hundredth = 26.67
   b. 129 to the nearest hundredth = .13
   c. 57.778 to the nearest hundredth = 57.78
3. a. 1.4285 to the nearest thousandth = 1.429
   b. 5.8333 to the nearest thousandth = 5.833
   c. .0194 to the nearest thousandth = .019

Lesson 26 (page 170)

1. .023   .0015   .03
2. .00016   .0327   .0055
3. 1.89   .002   .0019
4. .25   .0126   .4832
5. .013   5.4   .09
6. .004   .021   .0007
7. .23 meter
8. 1.25 pounds

Lesson 27 (pages 171-172)

1. a. apples
   b. 7.3 kg
   ```
   4)29.2
     28
     ──
     1 2
     1 2
     ───
       0
   ```

Note:
kg = kilogram

2. a. dollars
   b. dollars
   c. $ 7.65
   ```
   40)$306.00
      280
      ───
       26 0
       24 0
       ────
        2 00
        2 00
        ────
   ```

3. a. dollars
   b. dollars
   c. $ 1.04
   ```
   12.5.)$13.0.00
         12 5
         ────
            50
             0
         ────
          5 00
          5 00
          ────
   ```

4. a. miles
   b. three
   c. 83.3   to the nearest mile
   ```
   3)250.0    = 83 miles
     24
     ──
     10
      9
      ─
      1 0
        9
   ```

5. a. $990
   b.   20
   ```
   49.50.)990.00.
          990 0
          ─────
             00
   ```

Lesson 28 (pages 173-174)

6. a. price
   b. $ .183   to the nearest penny
   ```
   6.5.)1.1.900     = $.18
         6 5
         ───
         5 40
         5 20
         ────
           200
           195
   ```

1. a.   43.5
         53
       +26.5
       ─────
       123.0 pounds
   b. three
   c.   41 pounds
   ```
   3)123
   ```

2. a.   3.5
         4.5
        +5.5
        ────
        13.5 hours
   b. dollars
   c. dollars
   d. $ 4.20
   ```
   13.5.)$56.7.00
          54 0
          ────
           2 70
           2 70
           ────
             00
   ```

3. a. multiplication
   b. divide by 12
   c. $19,3 44
       ×   .25
       ────────
        96 7 20
       386 8 8
       ────────
       $483 6.00
   d. $ 403
   ```
   12)$4836
   ```
   e. $ 1,612
   ```
   12)$19,344
   ```
   f. $ 1,6 12
      ×   .25
      ───────
        8 0 60
       32 2 4
       ───────
       $40 3.00

4. a. 18
       16.7
      +14.5
       49.2 miles
   b. miles
   c. 16.4 miles
      3)49.2

Lesson 29 (page 175)

1. 4.7   3.5   inches
    3.3   3)10.5
   +2.5
    10.5

2. 86.2   207.8   miles
    306.9 3)623.4
   +230.3
    623.4

3. 7.25   6.25   pounds
    3.4   3)18.75
   +8.1
    18.75

4. 1.089 $1.119   to the nearest
   +1.149 2)$2.238   penny = $1.12
    $2.238

5.   74   80.6   to the nearest whole
     78   5)403.0   number = 81°
     83
     86
    +82
    403

6. .435   .3415   to the nearest
    .325 4)1.3660   thousandth = .342
    .316
   +.290
    1.366

Lesson 30 (pages 176-177)

1.   1 7.6   to the nearest whole
    8.5.)150.0.0   number = 18 mpg
          85
          65 0
          59 5
           5 50
           5 10

2.   11.1   to the nearest whole
    9)100.0   number = 11 mpg
      9
      10
      9
      1 0
      9

3.   1 3.6   to the nearest whole
    12.5.)170.0.0   number = 14 mpg
          125
          45 0
          37 5
           7 50
           7 50

4.   5.77   to the nearest tenth
    45)260.00   = 5.8 mpg
       225
        35 0
        31 5
         3 50
         3 15

5.   4.75   to the nearest tenth
    30.5.)145.0.00   = 4.8 mpg
          122 0
           23 0 0
           21 3 5
            1 6 50
            1 5 25

6.   8.57   to the nearest tenth
    35)300.00   = 8.6 mpg
       280
        20 0
        17 5
         2 50
         2 45

7.   9.2   to the nearest whole
    50)460.0   number = 9 mpg
       450
        10 0
        10 0

Answers 203

Lesson 31 (page 179)

Player	Average	
	ten-thousandths	thousandths
Randolph	.3250	.325
Gomez	.2586	.259
Smith	.4347	.435
Rigby	.2898	.290
Johnson	.3164	.316
Howard	.2054	.205
Acevedo	.1666	.167

DECIMALS REVIEW (page 180)

1. 1.3 2. .706 3. 7.14 4. 20.735
5. .6 6. .063 7. 4.94 8. 6.633
9. .28 10. 27.54 11. 5.733
12. .0152 13. .27 14. 1.4 15. .06
16. 600 17. .6 18. 5.04 19. .014
20. .3 21. .096 22. 4.2 23. .15
24. 7

25. 98.6
    +3.5
    ─────
    102.1

26. $3.00
    −1.77
    ─────
    $1.23 million

27. 37.5
     3.25
     2.5
    + 4.5
    ─────
     47.75

28. 7634.1   205.9
    −7428.2   +115
    ──────   ─────
     205.9   320.9 miles

29. 1983

30. 344.5 million
    −137.0
    ──────────────
     207.5 million

31. 167.0 million
    −125.2
    ──────────────
     41.8 million

32. 167.0   295.2
    +125.2   −292.2
    ─────   ──────
     292.2   3.0 million

33. 3.6 centimeters   34. 7.2 centimeters

35. 4.4   36. 7.1
    −3.6   −3.6
    ────   ────
    .8 centimeter   3.5 centimeters

37. $6.35
    ×   7.5
    ───────
      3175
     4445
    ───────
    $47.625   to the nearest penny = $47.63

38.   52   39.   4.25
     ×2.5   ×$.18
     ────   ─────
     26.0   34 00
     104   42 5
     ────   ──────
     103.0 miles $.76 50   to the
                      nearest penny = $.77

40. 2.54 × 10 = 25.4 centimeters

41. A = LW   42. C = πd
    A = 4.6 × 3.4   C = 3.14 × 12
    A = 15.64 square   C = 37.68 feet
        meters

43. A = πr²   44. $ 5.85
    A = 3.14 × 30   40)$234.00
        × 30   200
    A = 2826 square   34 0
        feet   32 0
                              2 00
                              2 00
                                 0

45.   5.2   4.25 pounds
      4.95 3)12.75
    + 2.6
    ─────
     12.75

46.   10.90   to the nearest tenth
    11)120.00   = 10.9 mpg
       11
       ──
       10
        0
       ──
       10 0
        9 9
       ────
         10

47.   .2916   to the nearest thousandth
    24)7.0000   = .292
       4 8
       ───
       2 20
       2 16
       ────
         40
         24
         ──
         160
         144

204 *Answers*

48.
```
     $ 1,865        $186 5
 12)$22,380         ×   .3
     12             $559.5 = $559.50
     ‾‾
     10 3
      9 6
        78
        72
        60
        60
         0
```

49.
```
         $   4.40
 7.5.)$33.0.00
       30 0
        3 0 0
        3 0 0
           00
```

50. 9.8 kilograms
4)39.2